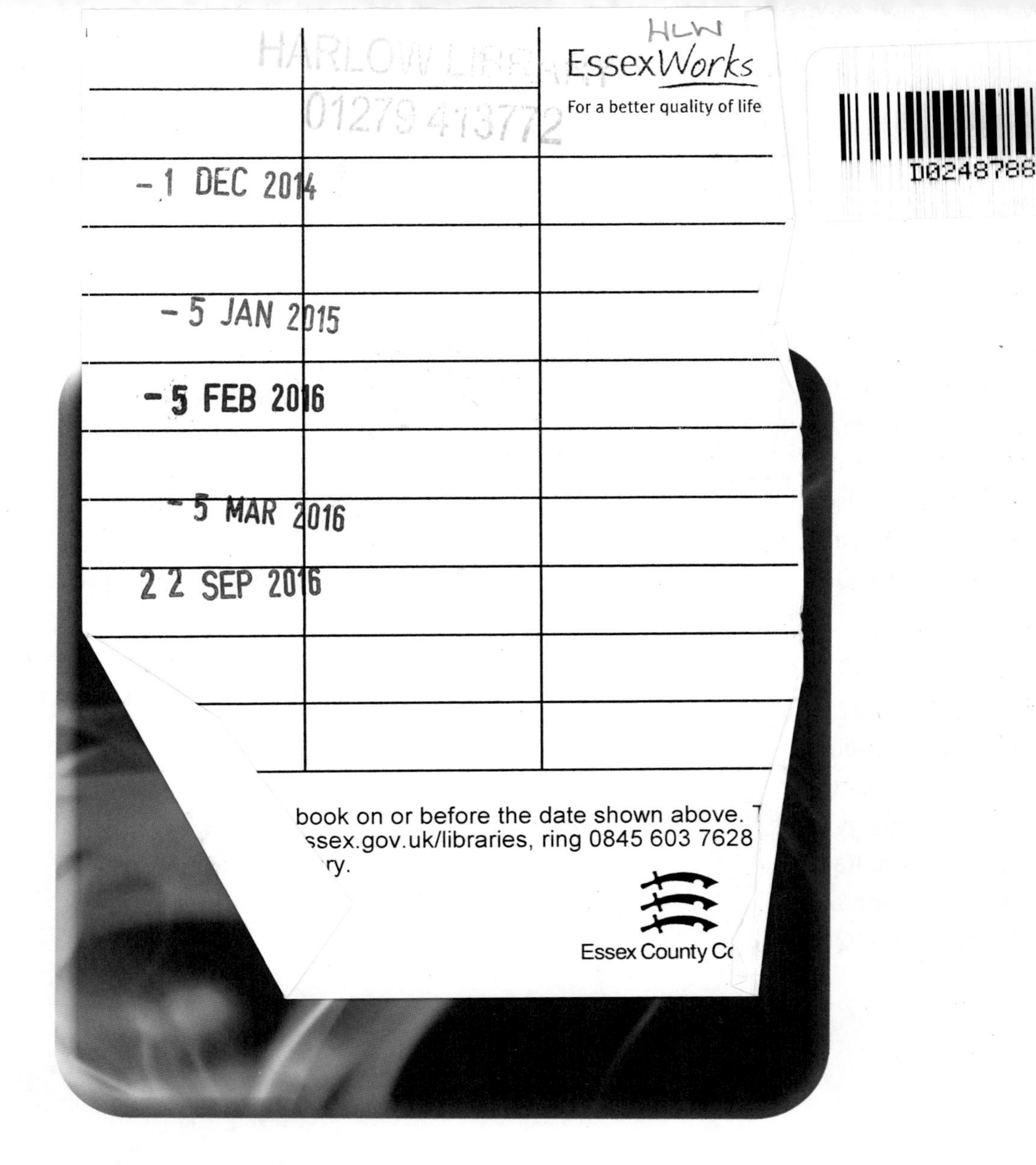

REVISION PLUS

AQA GCSE Physics

Revi[...]n Companion

Table of Contents

N.B. The numbers in brackets correspond to the reference numbers in the AQA Physics specification.

How Science Works Overview

The new AQA GCSE Physics specification incorporates two types of content:

- **Science Content** (example shown opposite). This includes all the scientific explanations and evidence that you need to be able to recall in your exams (objective tests or written exams). It is covered on pages 11–92 of the revision guide.
- **How Science Works** (example shown opposite). This is a set of key concepts, relevant to all areas of science. It is concerned with how scientific evidence is obtained and the effect it has on society. More specifically, it covers:
  - the relationship between scientific evidence and scientific explanations and theories
  - the practices and procedures used to collect scientific evidence
  - the reliability and validity of scientific evidence
  - the role of science in society and the impact it has on our lives
  - how decisions are made about the use of science and technology in different situations, and the factors affecting these decisions.

Because they are interlinked, your teacher may have taught the two types of content together in your science lessons. Likewise, the questions on your exam papers are likely to combine elements from both types of content. To answer them you will need to recall the relevant scientific facts and draw on your knowledge of how science works.

The key concepts from How Science Works are summarised in this section of the revision guide. You should be familiar with all of them, especially the practices and procedures used to collect scientific data (from all your practical investigations). Make sure you work through them all. Make a note of anything you are unsure about and then ask your teacher for clarification.

Methods We Use to Generate Electricity P1

P1.4 Methods we use to generate electricity

Various energy sources can be used to generate the electricity that we need. We need to understand the benefits and costs associated with each method to select the best one to use in a particular situation. To understand this, you need to know:

- how energy is produced
- about renewable and non-renewable energy sources
- the advantages and disadvantages of using different energy sources
- how electricity is distributed, including the role of transformers.

Non-renewable Energy Sources

Coal, oil and gas are energy sources that formed over millions of years from the remains of plants and animals. They are called **fossil fuels** and are used to produce most of the energy that we use. However, because they cannot be replaced within a lifetime they will eventually run out. They are therefore called **non-renewable** energy sources.

Nuclear fuels, such as uranium and plutonium, are also non-renewable. Nuclear fission is the splitting of a nucleus releasing neutrons that collide with other nuclei causing a chain reaction that generates huge amounts of heat energy. However, nuclear fuel is not burnt (like coal, oil or gas) to release energy and is not classed as a fossil fuel.

Generating Electricity from Non-renewable Sources

In power stations, fossil fuels are burned to release heat energy, which boils water to produce steam. The steam is used to drive turbines, which are attached to electrical generators. A gas jet turbine generator heats air instead of water giving it a quicker start-up time.

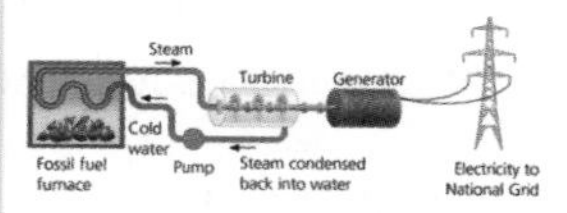

Nuclear fuel is used to generate electricity in a similar way. A reactor is used to generate heat by nuclear fission. A heat exchanger is used to transfer the heat energy from the reactor to the water, which turns to steam and drives the turbines.

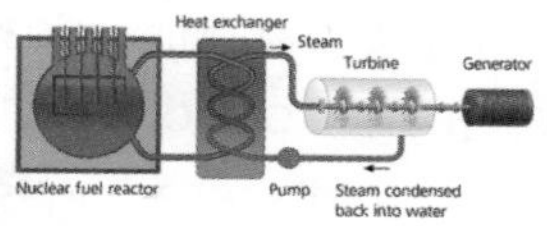

Biofuels

Biofuels – e.g. wood and rapeseed oil (used for bio-diesel) – are also burnt to release energy. Because the plants used grow relatively quickly, the fuels burned can be replaced. For this reason biofuels are classed as **renewable** energy sources. Plant and animal waste from farming can also be used as a biofuel.

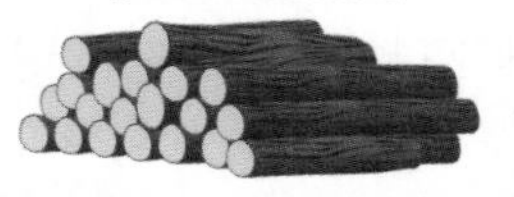

21

P2 How Science Works

You need to be able to construct distance–time graphs for a body when the body is stationary or moving in a straight line with constant speed.

Example

An athlete is training for a marathon. She runs at a constant speed for 1 minute and then rests for 20 seconds to allow her pulse rate to slow down. She repeats this four times. For each minute that she is running, she covers 400m. This information can be plotted on a distance–time graph.

The sloping lines show when the athlete is running and the flat lines show when she is stationary. The sloping lines are all at the same angle, showing that the athlete was always running at the same speed.

Don't forget to label both axes.

Plot each point carefully.

Make sure you use an appropriate scale.

You need to be able to construct velocity–time graphs for a body moving with a constant velocity or a constant acceleration.

Example

The same athlete practises a sprint finish by running in a straight line at a constant speed (i.e. velocity) of 7m/s for 20 seconds before gradually increasing her speed to a sprint. It takes her 20 seconds to reach a top speed of 9m/s accelerating at a constant rate. She then maintains this top speed for a further 20 seconds.

The flat lines show where the athlete was running at a constant velocity. The sloping line shows where she increased her velocity, accelerating at a constant rate. Graphs like this provide a clear visual representation of data, which can make it much easier to understand.

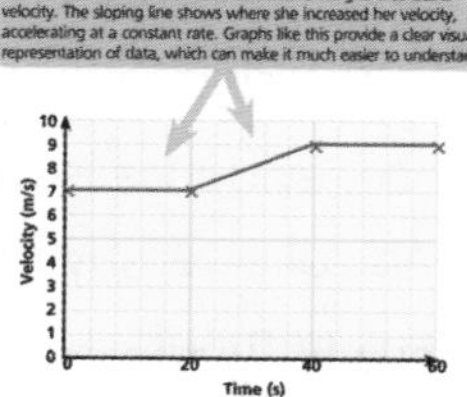

42

How Science Works Overview

How Science Works

The AQA GCSE Physics specification includes activities for each sub-section of science content. These require you to apply your knowledge of how science works and will help develop your skills when it comes to evaluating information, developing arguments and drawing conclusions.

These activities are dealt with on the How Science Works pages throughout the revision guide. Make sure you work through them all, because questions relating to the skills, ideas and issues covered on these pages could easily come up in the exam. Bear in mind that these pages are designed to provide a starting point from which you can begin to develop your own ideas and conclusions. They are not meant to be definitive or prescriptive.

Practical tips on how to evaluate information are included in this section, on page 9.

What is the Purpose of Science?

Science attempts to explain the world we live in. The role of a scientist is to collect evidence through investigations to:

- explain phenomena (i.e. explain how and why something happens)
- solve problems.

Scientific knowledge and understanding can lead to the development of new technologies (e.g. in medicine and industry), which have a huge impact on society and the environment.

Scientific Evidence

The purpose of evidence is to provide facts that answer a specific question, and therefore support or disprove an idea or theory.

In science, evidence is often based on data that has been collected by making observations and measurements.

A scientifically literate citizen should be equipped to question the evidence used in decision making. Evidence must therefore be approached with a critical eye. It is necessary to look closely at:

- how measurements have been made
- whether opinions drawn are based on valid and reproducible evidence rather than non-scientific ideas (prejudices and hearsay)
- whether the evidence is reproducible, i.e. it can be reproduced by others
- whether the evidence is valid, i.e. is reproducible, repeatable and answers the original question.

N.B. If data is not reproducible or repeatable, it cannot be valid.

To ensure scientific evidence is repeatable, reproducible and valid, scientists employ a range of ideas and practices that relate to the following:

1. **Observations** – how we observe the world.
2. **Investigations** – designing investigations so that patterns and relationships can be identified.
3. **Measurements** – making measurements by selecting and using instruments effectively.
4. **Presenting data** – presenting and representing data.
5. **Conclusions** – identifying patterns and relationships and making suitable conclusions.
5. **Evaluation** – considering the validity of data and appropriateness of methods used.

These six key ideas are covered in more detail on the following pages.

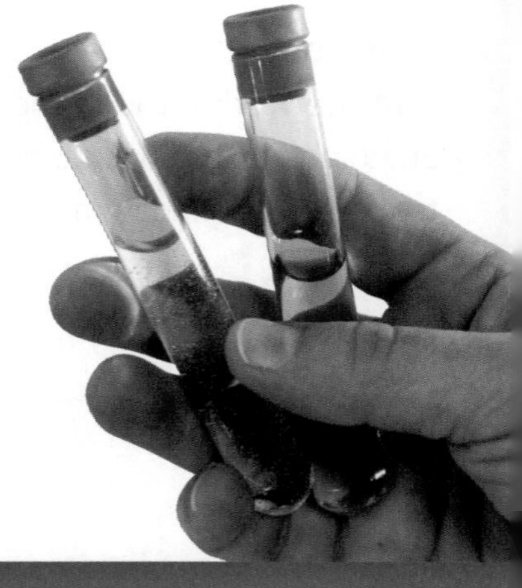

❶ Observations

Most scientific investigations begin with an observation, i.e. a scientist observes an event or phenomenon and decides to find out more about how and why it happens.

The first step is to develop a **hypothesis**, i.e. to suggest an explanation for the phenomenon. Hypotheses normally propose a relationship between two or more **variables** (factors that can change). Hypotheses are based on careful observations and existing scientific knowledge, and often include a bit of creative thinking.

The hypothesis is used to make a prediction, which can be tested through scientific investigation. The data collected during the investigation might support the hypothesis, show it to be untrue, or lead to the development of a new hypothesis.

Example

A biologist **observes** that freshwater shrimp are only found in certain parts of a stream.

The biologist uses current scientific knowledge of freshwater shrimp behaviour and water flow to develop a **hypothesis**, which relates the distribution of shrimp (first variable) to the rate of water flow (second variable).

Based on this hypothesis, the biologist **predicts** that shrimp can only be found in areas of the stream where the flow rate is beneath a certain value.

The prediction is **investigated** through a survey that looks for the presence of shrimp in different parts of the stream, which represent a range of different flow rates.

The **data** shows that shrimp are only present in parts of the stream where the flow rate is below a certain value (i.e. the data supports the hypothesis). However, it also shows that shrimp are not *always* present in parts of the stream where the flow rate is below this value.

As a result, the biologist realises that there must be another factor affecting the distribution of shrimp. So, he **refines his hypothesis** to relate the distribution of shrimp (first variable) to the concentration of oxygen in the water (second variable) in parts of the stream where there is a slow flow rate.

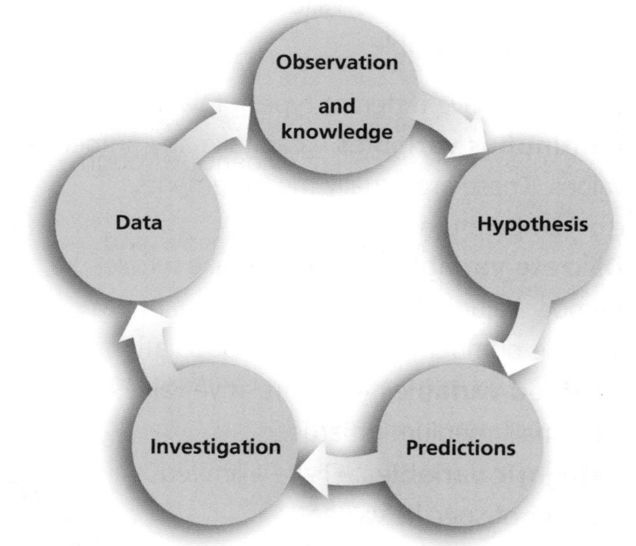

If new observations or data do not match existing explanations or theories (e.g. if unexpected behaviour is displayed) they need to be checked for reliability and validity.

In some cases it turns out that the new observations and data are valid, so existing theories and explanations have to be revised or amended. This is how scientific knowledge gradually grows and develops.

2 Investigations

An investigation involves collecting data to try to determine if there is a relationship between two variables. A variable is any factor that can take different values (i.e. any factor that can change). In an investigation there are three types of variables:

- **Independent variable**, which is changed or selected by the person carrying out the investigation. In the shrimp example on page 5, the independent variable is the flow rate of the water.
- **Dependent variable**, which is measured each time a change is made to the independent variable, to see if it also changes. In the shrimp example on page 5, the dependent variable is the number of shrimp.
- **Outside variables**, which are variables that could affect the dependent variable and must be controlled.

For a measurement to be valid it must measure only the appropriate variable.

Variables can have different types of values:

- **Continuous variables** – can take any numerical values. These are usually measurements, e.g. temperature or height.
- **Discrete variables** – can only take whole-number values. These are usually quantities, e.g. the number of shrimp in a population.
- **Ordered variables** – have relative values, e.g. small, medium or large.
- **Categoric variables** – have a limited number of specific values, e.g. the different breeds of dog: dalmatian, cocker spaniel, labrador, etc.

Numerical values tend to be more powerful and informative than ordered variables and categoric variables.

An investigation tries to establish if an observed link between two variables is one of the following:

- **Causal** – a change in one variable causes a change in the other, e.g. in a chemical reaction the rate of reaction (dependent variable) increases when the temperature of the reactants (independent variable) is increased.
- **Due to association** – the changes in the two variables are linked by a third variable. For example, a link between the change in pH of a stream (first variable) and a change in the number of different species found in the stream (second variable), may be the effect of a change in the concentration of atmospheric pollutants (third variable).
- **Due to 'chance occurence', i.e. coincidence** – the change in the two variables is unrelated; it is coincidental. For example, in the 1940s the number of deaths due to lung cancer increased and the amount of tar being used in road construction increased. However, one increase *did not* cause the other increase.

2 Investigations (cont)

Fair Tests

A fair test is one in which the only factor that can affect the dependent variable is the independent variable. Any other variables (outside variables) that could influence the results are kept the same.

A fair test is much easier to achieve in the laboratory than in the field, where conditions (e.g. weather) cannot always be physically controlled. The impact of outside variables, like the weather, has to be reduced by ensuring that all measurements are affected by the variable in the same way. For example, if you were investigating the effect of different fertilisers on the growth of tomato plants, all the plants would need to be grown in a place where they were subject to the same weather conditions.

If a survey is used to collect data, the impact of outside variables can be reduced by ensuring that the individuals in the sample are closely matched. For example, if you were investigating the effect of smoking on life expectancy, the individuals in the sample would all need to have a similar diet and lifestyle to ensure that those variables did not affect the results.

Control groups are often used in biological research. For example, in some drugs trials a placebo (a dummy pill containing no medicine) is taken by one group of volunteers (the control group) and the drug is taken by another group. By comparing the two groups, scientists can establish if the drug (the independent variable) is the only variable affecting the volunteers and therefore whether or not it is a fair test.

Selecting Values of Variables

Care is needed in selecting the values of variables to be recorded in an investigation. For example, if you were investigating the effect of fertilisers on plant growth, you would need a range of fertiliser concentration that would give a measurable range of growth. Too narrow a range of concentration may fail to give any noticeable difference in growth. A trial run often helps to identify appropriate values to be recorded, such as the number of repeated readings needed and their range and interval.

Repeatability, Reproducibility and Validity

Repeatability measures how consistent results are in a single experiment. For example, a student measures the time taken for magnesium to react with hydrochloric acid. She repeats the test five times and finds that for each test the reaction takes 45 seconds. We can say that repeatability is high and the results are accurate.

In practice, it is unlikely that all five tests will give exactly the same result and therefore the mean (average) of a set of repeated measurements is often calculated to overcome small variations and get a best estimate of a true result.

An accurate measurement is one that is close to the true value. The purpose of an investigation will determine how accurate the data collected needs to be. For example, measurements of blood alcohol levels must be accurate enough to determine if a person is legally fit to drive.

Reproducibility measures the ability of an experiment to produce results that are the same each time it is carried out. For example, a whole class of students carry out the experiment above with magnesium and hydrochloric acid. The results can be said to be reproducible, and therefore reliable, if all students' results are very close to each other.

Validity questions if the results obtained can be used to prove or disprove the original hypothesis. It must consider the design of the experiment, the extent to which variables have been controlled and the reliability of results. The data collected must be precise enough (i.e. to an appropriate number of decimal places) to form a valid conclusion.

3 Measurements

When making measurements, errors may occur that affect the repeatability, reproducibility and validity of the results. These may be due to:

- **Variables that have not been controlled** – these may be variables that are beyond control.
- **Human error** – when making measurements, random errors can occur due to a lapse in concentration. Random errors can also result from inconsistent application of a technique. Systematic (repeated) errors can result from an instrument not being calibrated correctly or repeatedly being misused, or from consistent misapplication of a technique.
- **The resolution of the instruments used** – the resolution of an instrument is determined by the smallest change in value it can detect. For example, the resolution of bathroom scales is insufficient to detect the changes in mass of a small baby, whereas the scales used by a midwife have a higher resolution. It is therefore important to select instruments with an appropriate resolution for the task.

There will always be some variation in the actual value of a variable no matter how hard we try to repeat an event. For example, if the same model parachute was dropped twice from the same height, in the same laboratory and how long it took to fall to the ground was timed, it is very unlikely both drops would take exactly the same time. However, the two results should be close to each other. Any anomalous values should be examined to try to identify the cause. If the result is a product of a poor measurement, it should be ignored.

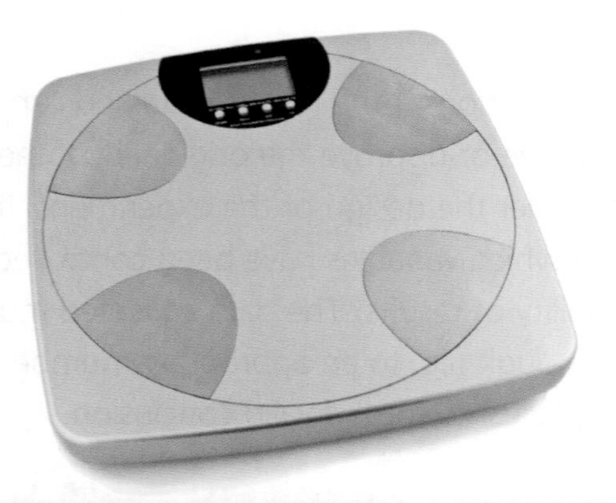

4 Presenting Data

When presenting data, two terms are frequently used:

- the **mean** (or average) – this is the sum of all the measurements divided by the number of measurements taken
- the **range** – this refers to the maximum and minimum values and the difference between them.

To explain the relationship between two or more variables, data can be presented in such a way that the pattern is more evident. The type of presentation used will depend on the type of variable represented.

Tables are an effective way to display data, but are limited in how much they can tell us about the design of an investigation.

Height (cm)	127	165	149	147	155	161	154	138	145
Shoe size	5	8	5	6	5	5	6	4	5

Bar charts are used to display data when one of the variables is categoric. They can also be used when one of the variables is discrete.

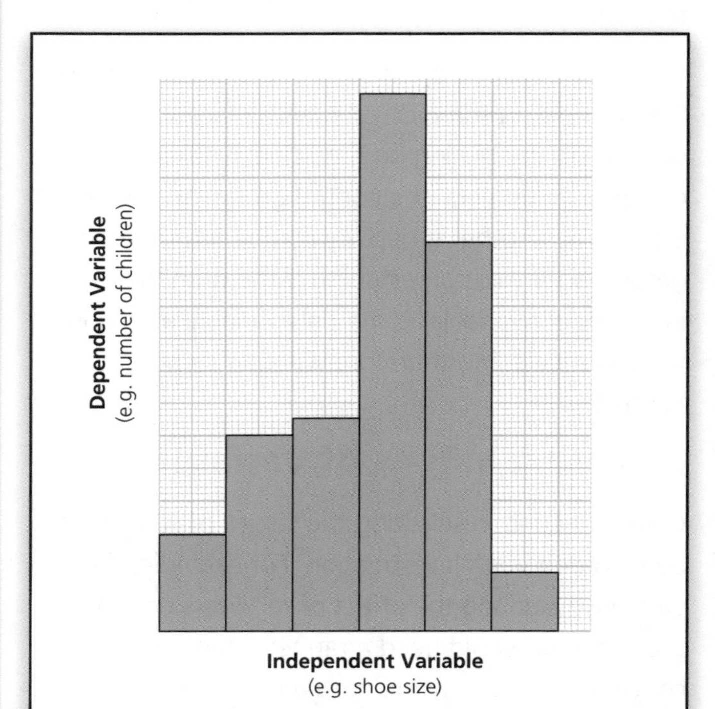

Scattergrams can be used to show an association between two variables. This can be made clearer by including a line of best fit.

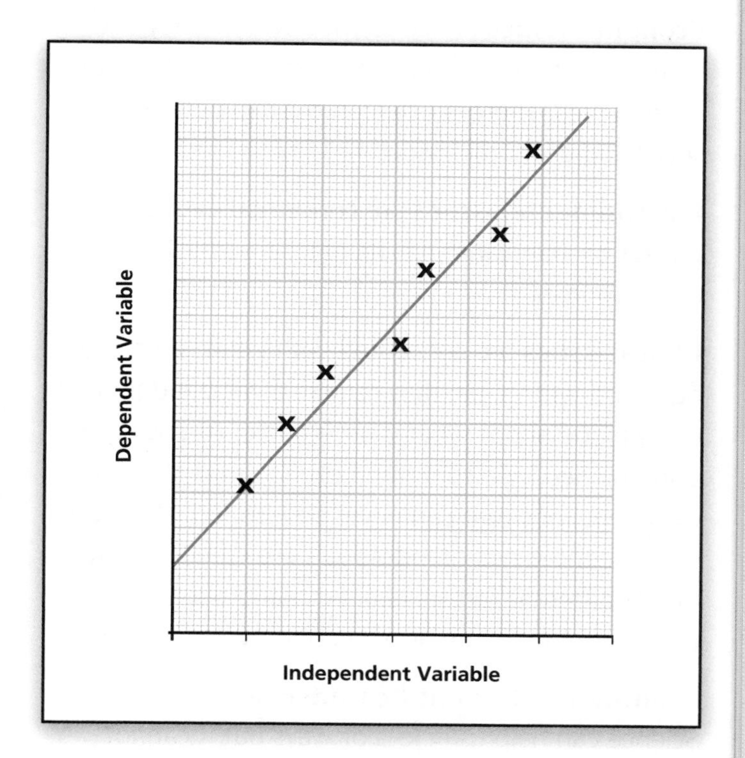

Line graphs are used when both the dependent and independent variables are continuous.

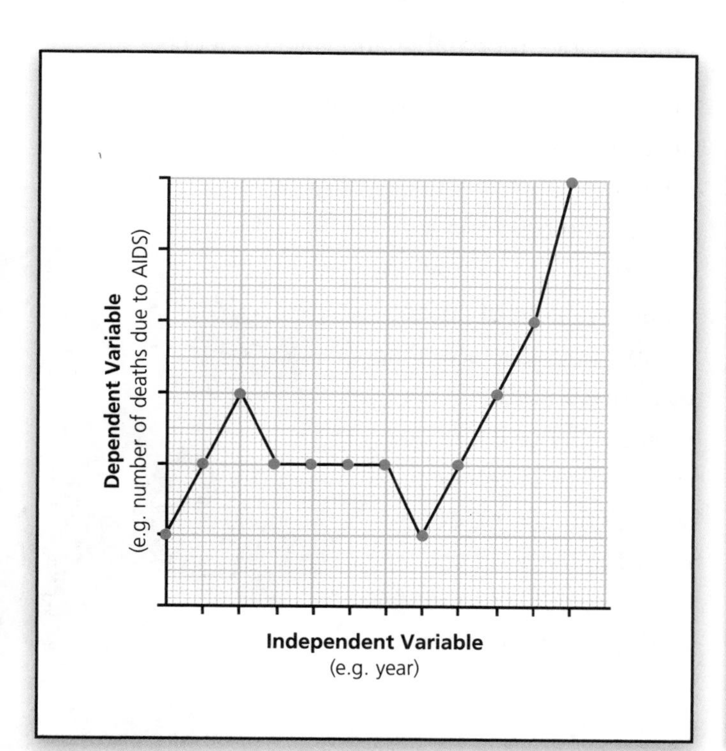

5 Conclusions

The patterns and relationships observed in the data represent what has happened in the investigation. However, it is necessary to look at the patterns and trends between the variables (bearing in mind limitations of the data), in order to draw conclusions.

Conclusions should:

- describe the patterns and relationships between variables
- take all the data into account
- make direct reference to the original hypothesis / prediction.

Conclusions should not:

- be influenced by anything other than the data collected
- disregard any data (other than anomalous values)
- include any speculation.

6 Evaluation

In evaluating a whole investigation, the repeatability, reproducibility and validity of the data obtained must be considered. This should take into account the original purpose of the investigation and the appropriateness of methods and techniques used in providing data to answer the original question.

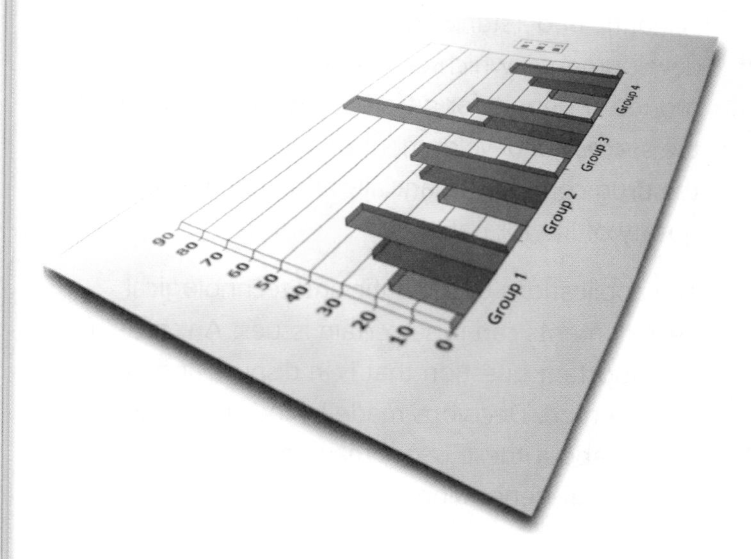

Societal Aspects of Scientific Evidence

Judgements and decisions relating to social-scientific issues may not be based on evidence alone. Sometimes other factors may have an influencing role. For example:

- **Bias** – evidence must be scrutinised for any potential bias on the part of the experimenter. Biased information might include incomplete evidence or may try to influence how you interpret the evidence. For example, this could happen if funding for the investigation came from a party with a vested interest, e.g. a drug company wanting to highlight the benefits of their new drug but downplay the side effects in order to increase sales.
- **Weight of evidence** – evidence can be given undue weight or be dismissed too lightly due to:
  - political significance – if the consequences of the evidence were likely to provoke public or political unrest or disquiet, the evidence may be downplayed
  - status of the experimenter – evidence is likely to be given more weight if it comes from someone with academic or professional status or who is considered to be an expert in that particular field.

Science and Society

Scientific understanding can lead to technological developments, which can be exploited by different groups of people for different reasons. For example, the successful development of a new drug benefits the drug company financially and improves the quality of life for patients.

The applications of scientific and technological developments can raise certain issues. An 'issue' is an important question that is in dispute and needs to be settled. Decisions made by individuals and society about these issues may not be based on scientific evidence alone.

Social issues are concerned with the impact on the human population of a community, city, country, or even the world.

Economic issues are concerned with money and related factors like employment and the distribution of resources. There is often an overlap between social and economic issues.

Environmental issues are concerned with the impact on the planet, i.e. its natural ecosystems and resources.

Ethical issues are concerned with what is morally right and wrong, i.e. they require a value judgement to be made about what is acceptable. As society is underpinned by a common belief system, there are certain actions that can never be justified. However, because the views of individuals are influenced by lots of different factors (e.g. faith and personal experience) there are also lots of grey areas.

Limitations of Scientific Evidence

Science can help us in lots of ways but it cannot supply all the answers. We are still finding out about things and developing our scientific knowledge. There are some questions that we cannot answer, maybe because we do not have enough reproducible, repeatable and valid evidence.

There are some questions that science cannot answer at all. These tend to be questions relating to ethical issues, where beliefs and opinions are important, or to situations where we cannot collect reliable and valid scientific evidence. In other words, science can often tell us *whether* something can be done or not and *how* it can be done, but it cannot tell us if it *should* be done.

P1.1 The transfer of energy by heating processes and the factors that affect the rate at which that energy is transferred

Thermal energy can be transferred by conduction, convection, infrared radiation, evaporation and condensation. You need to know which processes are important in particular situations and be able to explain how to reduce heat transfer in vacuum flasks, buildings, humans and animals. To understand this, you need to know:

- the role played by particles in heat transfer
- the main ways of insulating a home
- why silvered surfaces are used in vacuum flasks
- why trapped air is a good insulator.

Infrared Radiation

Thermal (infrared) **radiation** is the **transfer** of **thermal energy** (heat energy) by electromagnetic waves; no particles of matter are involved.

All objects emit and absorb thermal radiation. The hotter the object, the more energy it radiates. The amount of thermal radiation given out or taken in by an object depends on its surface, shape and dimensions.

An object will emit or absorb energy faster if there is a big difference in temperature between it and its surroundings. The rate of heat transfer can be slowed down by the use of insulation, which provides a barrier.

Under similar conditions, different materials transfer heat at different rates. At the same temperature:

- dark matt surfaces emit more radiation than light shiny surfaces
- dark matt surfaces absorb more radiation than light shiny surfaces because shiny surfaces are good reflectors of infrared radiation.

Kinetic Theory

The particles in solids, liquids and gases have different amounts of energy.

In a solid the particles have the least energy. They are not able to move but do vibrate around a fixed point.	
Heating a solid gives the particles enough energy to move around (although they cannot move far apart) and it melts to become a liquid.	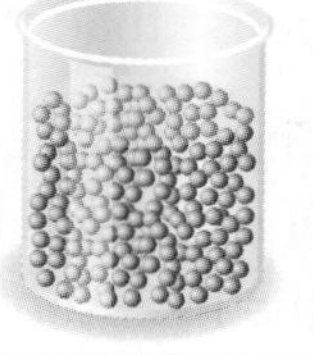
Further heating gives the particles enough energy to move very quickly. They separate from each other and the liquid evaporates to become a gas.	

Conduction

Conduction is the transfer of heat energy without the substance itself moving.

The structure of metals makes them good conductors of heat. As a metal becomes hotter, its tightly packed particles gain more kinetic energy and vibrate. This energy is transferred to cooler parts of the metal by delocalised electrons, which move freely through the metal, colliding with particles and other electrons.

N.B. An electron is a subatomic particle.

Conduction also occurs in non-metal solids because the particles can pass energy from one to the next by vibration. However, the lack of free electrons makes most non-metals poor conductors. Gases are bad conductors because the particles are so far apart.

Heat energy is conducted up the poker as the hotter parts transfer energy to the colder parts

Convection

Convection is the transfer of heat energy through movement. This occurs in liquids and gases and creates convection currents.

In a liquid or gas the particles nearest the heat source move faster and become further apart. This causes the substance to expand and become less dense than the colder parts.

The warm liquid or gas rises up and colder, denser liquid or gas moves into the space created (close to the heat source).

Example 1

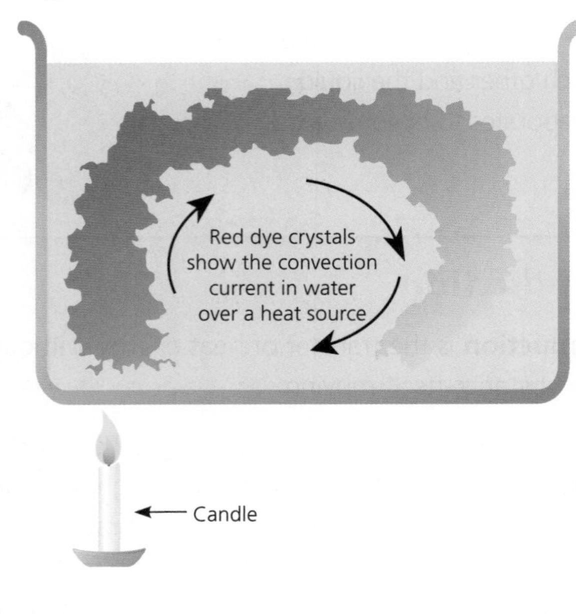

Example 2

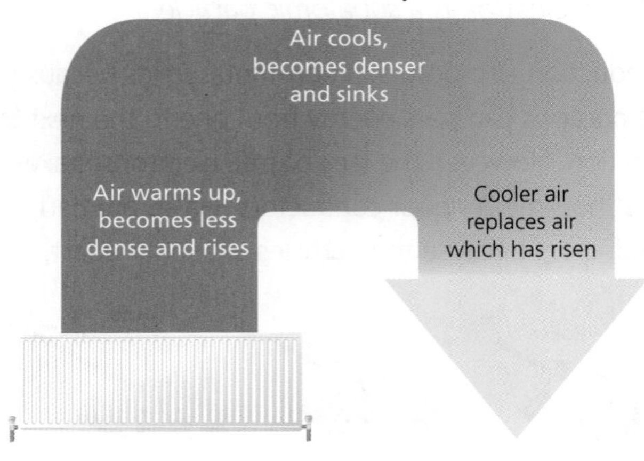

Condensation

The particles in a gas have more energy than those in a liquid. When a gas condenses to form a liquid, this energy is released and can make the temperature increase.

Condensation can occur when a warm gas comes in contact with a cold surface, e.g. water vapour in breath condensing on a window.

The surface needs to be cold enough to cool the particles so that they no longer have enough energy to move around quickly as in a gas. The colder the surface the greater the rate of condensation.

Evaporation

A liquid evaporates when its particles have gained enough energy to escape the surface of the liquid and become a gas.

For example, you feel cold when you get out of a swimming pool because the water is evaporating from your body and in doing so it takes heat energy from you.

Clothes on a washing line dry quickest on a sunny and windy day because the Sun provides the energy for the particles to escape from the surface, and the wind blows them away (making space for more water particles to escape).

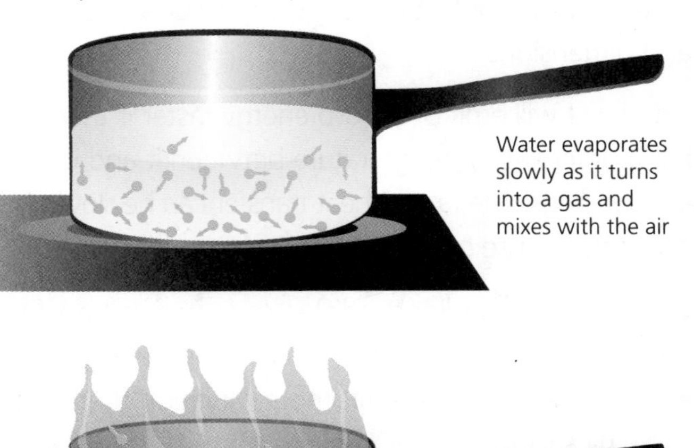

You need to be able to apply ideas about energy transfer to keeping warm or cool.

The rate at which heat is transferred depends on a number of factors:

1. A large surface area compared to volume will gain and lose heat quicker.
   Example – desert foxes have big, thin ears to lose heat quickly, but arctic foxes have smaller, fatter ears to minimise heat loss.

2. Different materials transfer heat at different rates.
   Example – fur, feathers and human clothing are all poor conductors used to reduce heat loss. In fact, most importantly, they all trap air (a bad conductor), which reduces heat loss even more.
3. Humans sweat and dogs pant.
   Example – as the moisture on our skin, or in a dog's mouth, evaporates it takes heat from its surroundings. This keeps us, and the dog, cool.
4. The surface the object is in contact with.
   Example – when you stand on a carpet in bare feet the floor feels warm because it traps air, which is an insulator. However, a tiled floor is a better conductor so it feels colder even if it is at the same temperature as the carpet.
5. The bigger the temperature difference between an object and its surroundings the faster it transfers heat.
   Example – during a cold winter a house costs more to heat compared to a mild winter. The bigger the temperature difference between inside and outside the quicker the house loses heat.

Examples

A **car radiator** is black, therefore it radiates heat well, and also has a large surface area due to thin cooling fins.

A **vacuum flask** greatly reduces heat flow from the inside to the outside or vice versa. This means that it will keep hot drinks hot or cold drinks cold. A flask is normally made from plastic, which is a poor conductor. The shiny (silvered) sides reflect infrared radiation and stop heat transfer. The vacuum contains no particles, so neither conduction nor convection can take place. The screw top prevents evaporation from the surface and convection currents at the top.

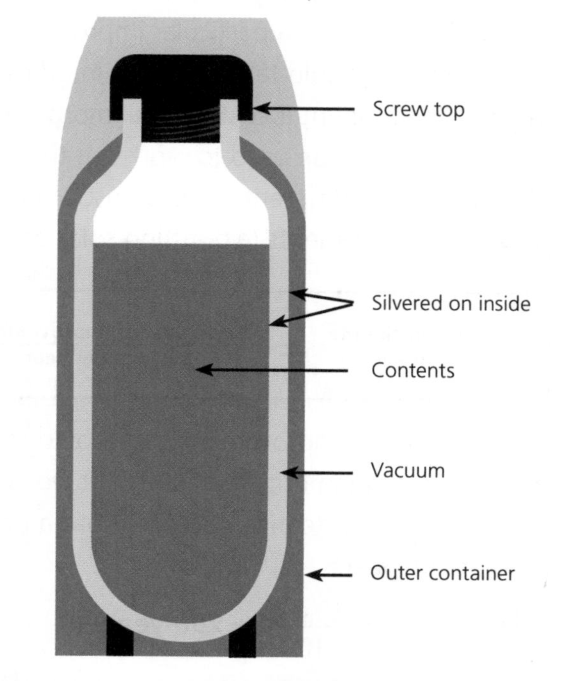

Heating and Insulating Buildings

U-values

The *U*-value of a material indicates how effective a material is as an insulator because it shows how quickly heat energy can pass through. A low *U*-value means that heat flows through it slowly. Therefore, the lower the *U*-value the better the material is at insulating.

Loft insulation

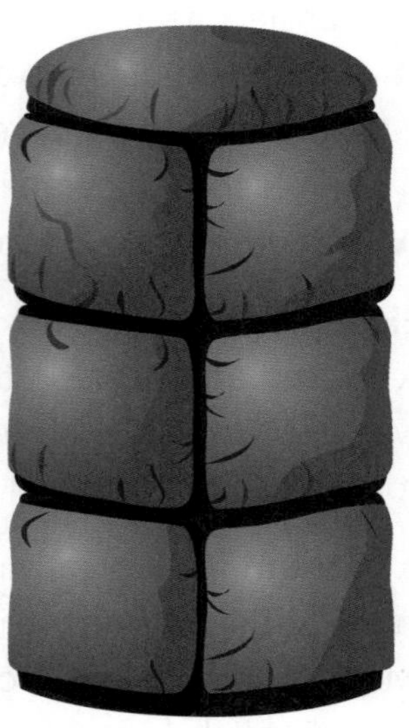

Boiler jacket

Efficiency and Payback Time

When designing houses it is important to consider the *U*-value of the materials used and compare the insulating benefits against the cost of the material. The payback time of a particular improvement tells us how long it would take to make, in efficiency savings, the amount it cost for the improvement. Payback time can be used to work out the cost effectiveness of different types of insulation as well as other improvements (e.g. fitting solar panels).

$$\text{Payback time (years)} = \frac{\text{Total cost of improvement}}{\text{Savings per year}}$$

For example, an improvement that would save £100 per year in heating but costs £20 000 would not be considered cost effective because it has a payback time of 200 years.

Payback time $= \frac{20\ 000}{100} =$ **200 years**.

Specific Heat Capacity

The specific heat capacity of a substance is the amount of energy needed to change the temperature of one kilogram of it by one degree Celsius.

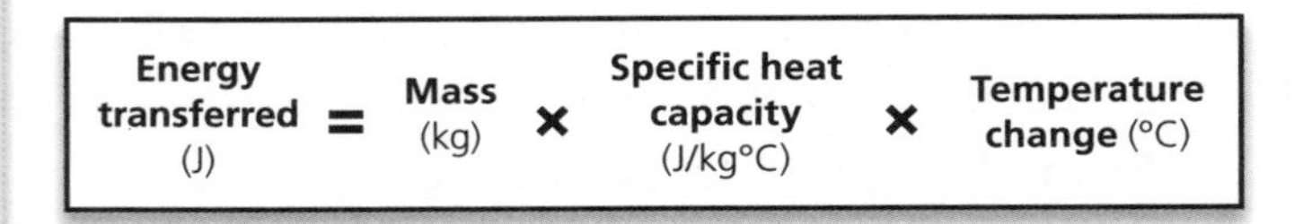

$$\text{Energy transferred (J)} = \text{Mass (kg)} \times \text{Specific heat capacity (J/kg°C)} \times \text{Temperature change (°C)}$$

This is useful for deciding which materials to use in heating and cooling applications.

For example, water has a high specific heat capacity, which means it can store a lot of heat energy without getting too warm. This makes it useful as a coolant.

Water is also used in solar heating panels. Water-filled panels placed on a roof absorb heat from the Sun. This radiation energy warms the water. The water in the panel stores a lot of heat energy, which can then be used to heat buildings or provide domestic hot water.

Solar panels are often black and have a large surface area to absorb as much infrared energy as possible.

You need to be able to compare ways in which energy is transferred in and out of objects by heating and ways in which the rate of transfer can be varied.

Every year, people in the UK spend more money on energy (gas and electricity) to heat their homes than they need to. This is because energy gets lost and wasted.

Percentage of Total Heat Loss

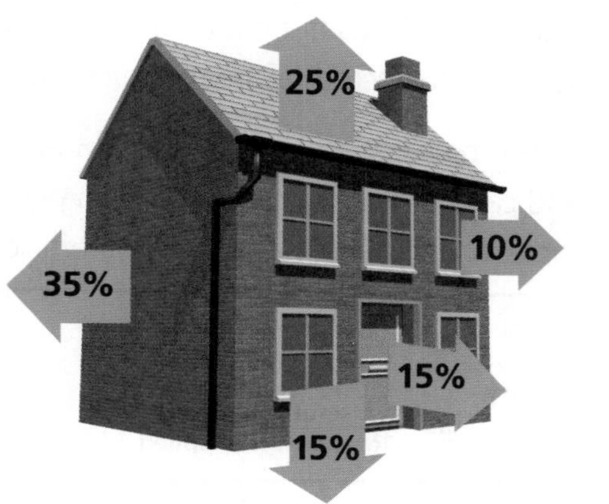

For example, an uninsulated house can lose up to 75% more heat than an insulated house.

Heat energy is transferred from homes into the environment by:

- **conduction** – through the walls, floor, roof and windows
- **convection** – convection currents coupled with cold draughts from gaps in doors and windows cause heat energy carried by warm air to rise up to the roof space where it is easily lost
- **radiation** – from the surface of the walls, roof and through the windows.

Insulating a house will help it to retain the heat in winter, and also help to keep it cool in the summer. The table below outlines how heat can be lost and how this loss can be reduced.

Where Heat is Lost	Preventative Measure	Benefits	Problems
Roof	• Roof insulation – traps a layer of air between fibres or insulating material.	• Can reduce heat loss by 20–25%. • Many different methods to suit all homes. • Short payback time.	• Requires suitable safety precautions to be taken, e.g. wearing dust mask, gloves, etc.
Under doors and windows	• Draught excluders – keep as much warm air inside as possible.	• Can reduce heat loss by up to 15%. • Cheap and easy to install. • Short payback time.	• Must make sure that air vents are not blocked – fresh air needs to circulate to prevent dry rot.
Walls	• Cavity wall insulation and internal thermal boards.	• Can reduce heat loss by 35%.	• Expensive. • Long payback time.
Windows	• Double glazing – traps air between two sheets of glass. • Curtains – stop heat loss through convection.	• Double glazing can reduce heat loss by up to 10%. • Curtains are cheap and easy to install.	• Double glazing is expensive and has a long payback time.
Floor	• Carpets, rugs and underfloor insulation can help to stop heat loss through the floor.	• Carpets and rugs are easy to install.	• Underfloor insulation is expensive and has a long payback time.

P1.2 Energy and efficiency

Energy can be transferred usefully, stored or dissipated, but cannot be created or destroyed. When appliances transfer energy some is always dissipated to the surroundings (wasted). To understand energy efficiency, you need to know:

- how energy is transferred by a range of appliances
- how to calculate efficiency
- how to interpret and draw Sankey diagrams.

Transferring Energy

When devices transfer energy, only part of the energy is usefully transferred to where it is wanted. The rest of the energy is 'wasted'. Wasted energy is eventually transferred to the surroundings, which become warmer. As energy is dissipated to the surroundings it becomes increasingly spread out and so becomes less useful.

Replacing old technology with newer more efficient technology will mean that less of the energy supplied is wasted. When making decisions we need to look at how cost effective the replacement is. The **payback time** of replacing an appliance with a more efficient one depends on how much it costs and how much it saves. When calculating this payback time we use exactly the same method as for insulation.

Efficiency

The greater the proportion of energy that is usefully transferred, the more **efficient** we say the device is. For example:

- A car engine is 30% efficient – much more energy is wasted (in heat and sound), than is transferred into useful kinetic energy.
- A microwave oven is 60% efficient – more energy is transferred into useful heat, light and kinetic energy than is wasted (as heat and sound).

Efficiency can be calculated using either energy or power and can be given as a decimal or a percentage. For example, an efficiency of 0.2 is 20% efficient.

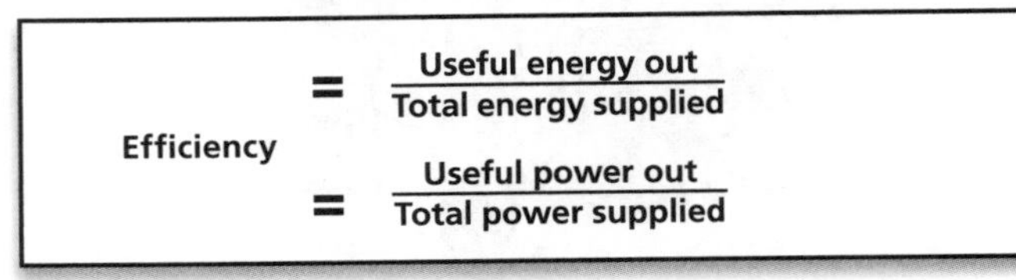

For example, if a 100W standard light bulb has a light output of 5W, the efficiency would be:

Energy Efficiency of a Television

Electrical energy 200 joules/sec

Wasted energy Heat 150 joules/sec

Useful energy Light 20 joules/sec

Useful energy Sound 30 joules/sec

Only a quarter of the energy supplied to a television is usefully transferred into light and sound. Therefore it is only 25% efficient.

Sankey Diagrams

Sankey diagrams are used to illustrate energy transfers. They provide a visual representation of how much energy there is of each type. The widths of the arrows are proportional to the amount of energy they represent.

The Sankey diagram below illustrates a standard light bulb. The diagram shows that most of the input electrical energy is wasted as heat. When drawing a Sankey diagram, the useful energy is drawn going straight through and the wasted energy branching off.

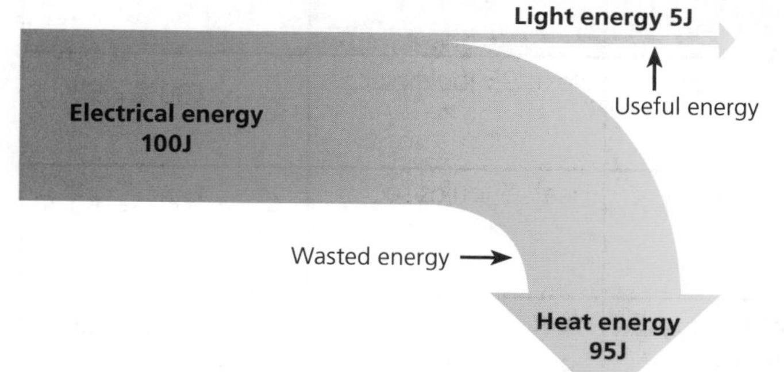

Example

In the example below, the Sankey diagram has been drawn on a square grid. It illustrates the energy transfer that takes place in a car engine, which is 30% efficient.

In the diagram the input energy is 10 squares wide and the useful energy is 3 squares wide.

$\frac{3}{10}$ = 0.3 (or 30% efficient)

If the car engine uses 1000J of chemical energy it would usefully transfer 300J to kinetic energy. The rest would be wasted as 500J of heat and 200J of sound.

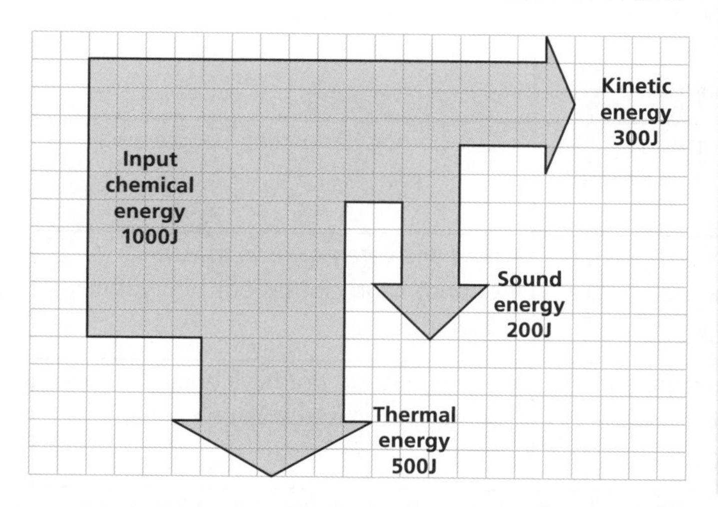

Remember that energy cannot be created or destroyed so when using Sankey diagrams the total energy output (useful + wasted) must **equal** the total energy input.

Example

A microwave oven has an input of 1000J. Draw a Sankey diagram to show the energy transfer of the oven (use a grid where 1 square represents 200J.) How many squares show the useful energy? What is the efficiency of the microwave oven?

The energy of the different outputs are as follows:

Useful: Kinetic – 1 square = 1 × 200 = 200J
Thermal – 1.5 squares = 1.5 × 200 = 300J
Light – 0.5 squares = 0.5 × 200 = 100J

Wasted: Sound – 1 square = 200J
Thermal – 1 square = 200J

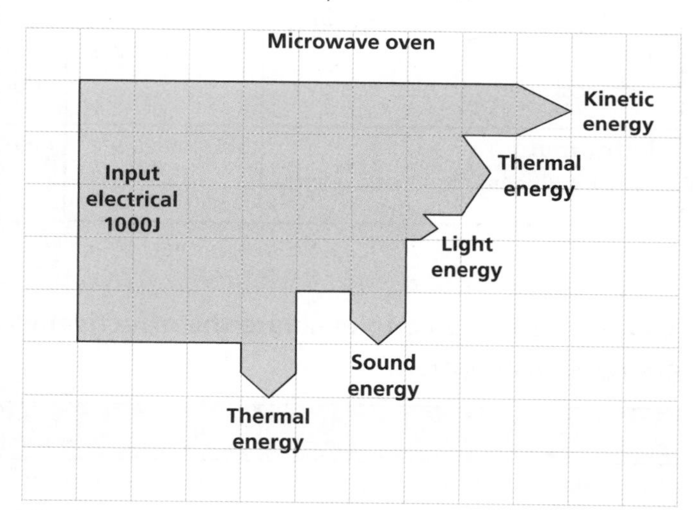

The total useful energy is **3 squares**.

Efficiency = $\frac{3}{5}$ = **0.6** (or 60% efficient)

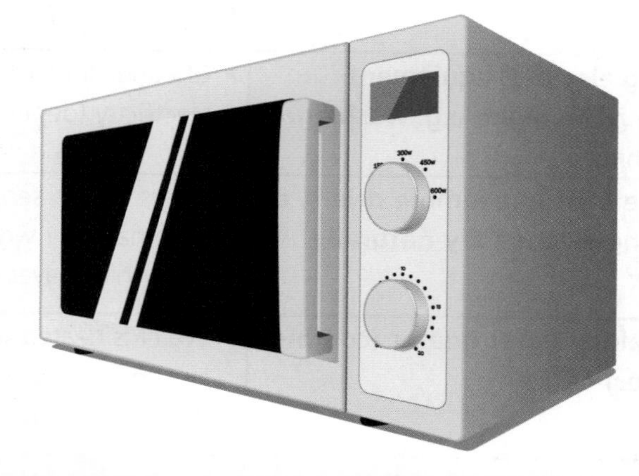

You need to be able to describe the energy transfers and the main energy wastage that occurs with a range of devices, and calculate the efficiency of a device using:

$$\text{Efficiency} = \frac{\text{Useful energy transferred by device}}{\text{Total energy supplied to device}} \times 100$$

Example

Some examples are shown in the table below.

Name of Device and Intended Energy Transfer	Energy In	Energy Out		Efficiency
		Useful	Wasted	
Standard light bulb – electrical energy to light energy	100 joules/sec	Light: 5 joules/sec	Heat: 95 joules/sec	$\frac{5}{100} \times 100\% =$ **5%**
Low energy (compact fluorescent) light bulb – electrical energy to light energy	25 joules/sec	Light: 20 joules/sec	Heat: 5 joules/sec	$\frac{20}{25} \times 100\% =$ **80%**
Kettle – electrical energy to heat energy	2000 joules/sec	Heat (in water): 1800 joules/sec	Heat: 100 joules/sec Sound: 100 joules/sec	$\frac{1800}{2000} \times 100\% =$ **90%**
Electric motor – electrical energy to kinetic energy	500 joules/sec	Kinetic: 300 joules/sec	Heat: 100 joules/sec Sound: 100 joules/sec	$\frac{300}{500} \times 100\% =$ **60%**

You need to be able to evaluate the effectiveness and cost effectiveness of methods used to reduce energy consumption.

Method	Benefits	Problems
Switching lights off when leaving a room	• Easy and simple way to reduce energy consumption.	• People forget to switch lights off or do not like being in a dark house.
Energy-efficient (compact fluorescent) light bulbs	• Use less power so cost less to run. • Less wasted energy. • Last longer than standard bulbs.	• More expensive to buy.
Using electrical equipment during the night, e.g. washing machine	• Cheaper time of day for using electricity (depending on the tariff).	• Can be noisy and keep people awake.
More efficient tumble driers, or letting clothes dry naturally	• Driers with a sensor stop automatically when the clothes are dry, which saves energy.	• New driers are expensive to buy. • Cannot hang clothes out to dry if it is raining
Tankless water heater (combi-boiler)	• Water is heated when needed so less energy is used to heat unnecessary water and keep it heated.	• Some units do not have enough power to supply to more than one tap at a time.

P1.3 The usefulness of electrical appliances

Electrical energy can be transferred easily across large distances and therefore is very suitable to use in our homes and industries. To understand this, you need to know:

- what energy transformations electrical devices bring about
- how energy and power are measured

Energy Transformation

Most of the energy transferred to homes and industry is electrical energy. Electrical energy is easily transformed to:

- heat (thermal) energy, e.g. an electric fire
- light energy, e.g. a lamp
- sound energy, e.g. stereo speakers
- movement (kinetic) energy, e.g. an electric whisk.

The **power** of an appliance is measured in **watts** (**W**) or **kilowatts** (**kW**). **Energy** is normally measured in **joules** (**J**). *N.B. 1W = 1J/s*

$$\textbf{Power} \text{ (W)} = \frac{\textbf{Energy} \text{ (J)}}{\textbf{Time} \text{ (s)}}$$

Calculating the Amount and Cost of Energy Transferred

The amount of energy transferred from the mains can be calculated using the equation:

$$\textbf{Energy transferred} \text{ (kilowatt-hour, kWh)} = \textbf{Power} \text{ (kilowatt, kW)} \times \textbf{Time} \text{ (hour, h)}$$

The **cost of energy** transferred from the mains can be calculated using the equation:

$$\textbf{Total cost} = \textbf{Number of kilowatt hours} \times \textbf{Cost per kilowatt hour}$$

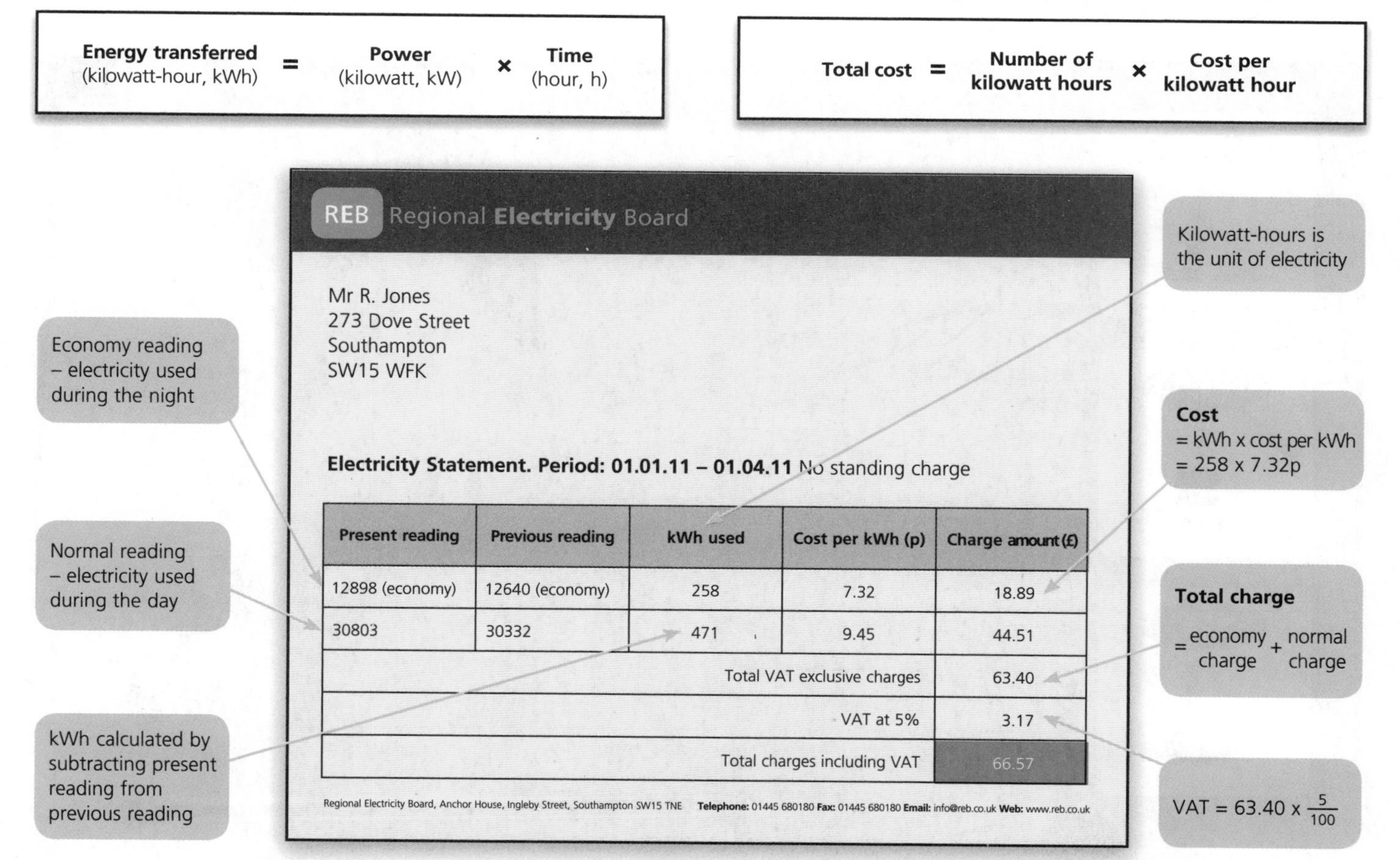

REB Regional **Electricity** Board

Mr R. Jones
273 Dove Street
Southampton
SW15 WFK

Electricity Statement. Period: 01.01.11 – 01.04.11 No standing charge

Present reading	Previous reading	kWh used	Cost per kWh (p)	Charge amount (£)
12898 (economy)	12640 (economy)	258	7.32	18.89
30803	30332	471	9.45	44.51
			Total VAT exclusive charges	63.40
			VAT at 5%	3.17
			Total charges including VAT	66.57

Regional Electricity Board, Anchor House, Ingleby Street, Southampton SW15 TNE **Telephone:** 01445 680180 **Fax:** 01445 680180 **Email:** info@reb.co.uk **Web:** www.reb.co.uk

You need to be able to compare and contrast different electrical devices that can be used for a particular application, and calculate the amount of energy transferred from the mains using this equation:

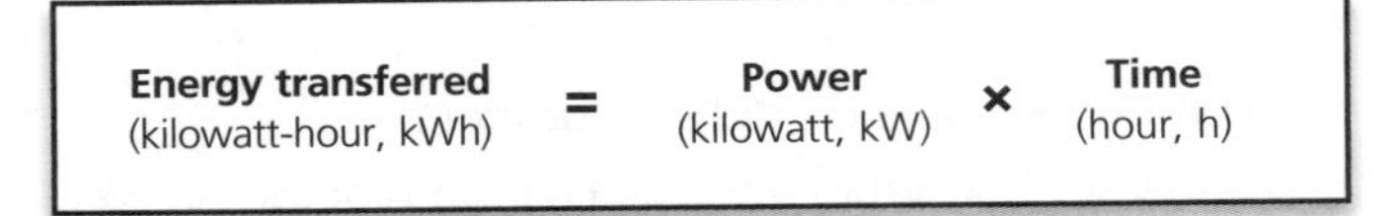

$$\text{Energy transferred (kilowatt-hour, kWh)} = \text{Power (kilowatt, kW)} \times \text{Time (hour, h)}$$

Example

Mr and Mrs Jones are shopping for a new shower for their house. They see three different showers advertised (see below).

They work out the advantages and disadvantages of each model in order to decide which one will be the most suitable for their needs. They also need to work out how much energy will be needed to power each one. They have calculated that every morning they each have a shower lasting, on average, 12 minutes, while their young son, Nick, has a shorter shower lasting 6 minutes (total time = 12 + 12 + 6 = 30 min or 0.5 hours).

Using the formula they calculate how much energy will need to be transferred from the mains to power each shower:

Power × Time = Energy

7kW: $7 \times 0.5 = 3.5$kWh

8.5kW: $8.5 \times 0.5 = 4.25$kWh

9.5kW: $9.5 \times 0.5 = 4.75$kWh

To compare the showers they work out the following:

- The 7kW shower will require the least amount of energy from the mains, but it has a less powerful spray and fewer options.
- The 8.5kW shower will require more energy than the 7kW shower, but it has a more powerful spray.
- The 9.5kW shower is more expensive to buy and requires more energy, but it has a much more powerful spray, more functions and can be connected to a cold water tank. This will be more useful for a family because they won't have to wait for the water to heat up.

7kW standard shower
- medium power jet
- suitable for any household
- choice of three colours

8.5kW electric shower
- three function power selector
- 21% more power than a standard 7kW shower
- fully temperature stabilised for maximum comfort

9.5kW supreme power shower
- connects to cold water supply
- 36% more power than average 7kW shower
- great for large families
- triple function – low, medium and high jet
- multi-spray shower head with massage function

P1.4 Methods we use to generate electricity

Various energy sources can be used to generate the electricity that we need. We need to understand the benefits and costs associated with each method to select the best one to use in a particular situation. To understand this, you need to know:

- how energy is produced
- about renewable and non-renewable energy sources
- the advantages and disadvantages of using different energy sources
- how electricity is distributed, including the role of transformers.

Non-renewable Energy Sources

Coal, oil and gas are energy sources that formed over millions of years from the remains of plants and animals. They are called **fossil fuels** and are used to produce most of the energy that we use. However, because they cannot be replaced within a lifetime they will eventually run out. They are therefore called **non-renewable** energy sources.

Nuclear fuels, such as uranium and plutonium, are also non-renewable. Nuclear fission is the splitting of a nucleus releasing neutrons that collide with other nuclei causing a chain reaction that generates huge amounts of heat energy. However, nuclear fuel is not burnt (like coal, oil or gas) to release energy and is not classed as a fossil fuel.

Generating Electricity from Non-renewable Sources

In power stations, fossil fuels are burned to release heat energy, which boils water to produce steam. The steam is used to drive turbines, which are attached to electrical generators. A gas jet turbine generator heats air instead of water giving it a quicker start-up time.

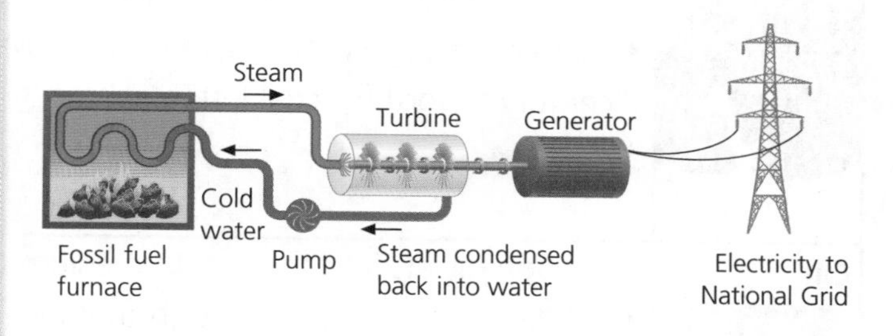

Nuclear fuel is used to generate electricity in a similar way. A reactor is used to generate heat by nuclear fission. A heat exchanger is used to transfer the heat energy from the reactor to the water, which turns to steam and drives the turbines.

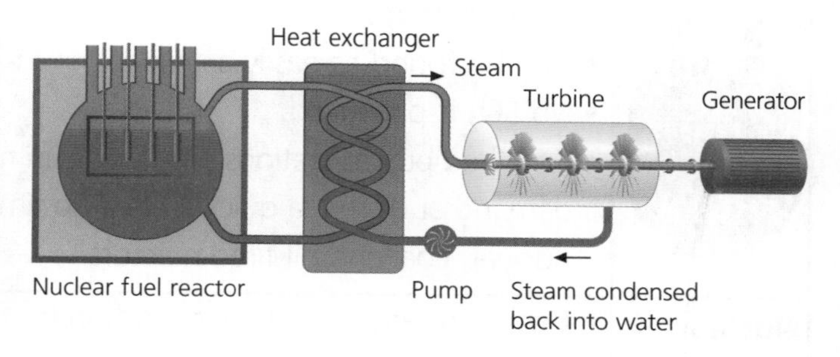

Biofuels

Biofuels – e.g. wood and rapeseed oil (used for bio-diesel) – are also burnt to release energy. Because the plants used grow relatively quickly, the fuels burned can be replaced. For this reason biofuels are classed as **renewable** energy sources. Plant and animal waste from farming can also be used as a biofuel.

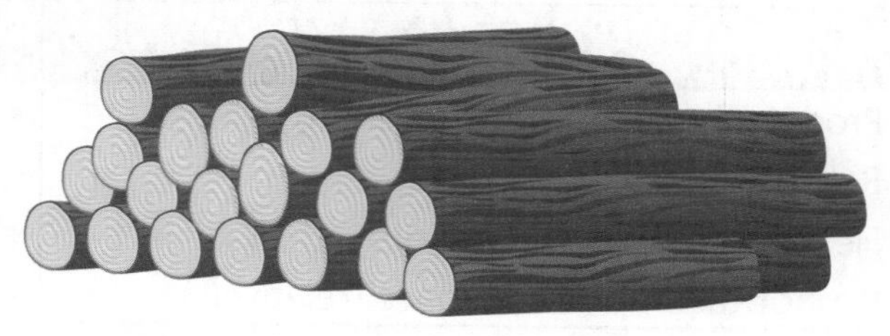

Comparing Non-renewable Sources of Energy

The energy sources below are used to provide most of the electricity we need in this country through power stations. Some of the advantages and disadvantages of each are listed below.

Source	Advantages	Disadvantages
Coal	• Relatively cheap and easy to obtain. • Coal-fired power stations are flexible in meeting demand and have a quicker start-up time than their nuclear equivalents. • Estimates suggest that there may be over a century's worth of coal left.	• Burning produces carbon dioxide (CO_2) and sulfur dioxide (SO_2). • Produces more CO_2 per unit of energy than oil or gas does. (CO_2 causes global warming.) • SO_2 causes acid rain unless the sulfur is removed before burning or the SO_2 is removed from the waste gases. Both of these add to the cost.
Oil	• Enough oil left for the short to medium term. • Relatively easy to find, though the price is variable. • Oil-fired power stations are flexible in meeting demand and have a quicker start-up time than both nuclear-powered and coal-fired reactors.	• Burning produces CO_2 and SO_2. (CO_2 causes global warming, SO_2 causes acid rain.) • Produces more CO_2 than gas per unit of energy. • Often carried between continents on tankers leading to the risk of spillage and pollution.
Gas	• Enough natural gas left for the short to medium term. • Can be found as easily as oil. • No SO_2 is produced. • Gas-fired power stations are flexible in meeting demand and have a quicker start-up time than nuclear, coal and oil-fired reactors.	• Burning produces CO_2, although it produces less than coal and oil per unit of energy. (CO_2 causes global warming.) • Expensive pipelines and networks are often required to transport it to the point of use.
Nuclear	• Cost and rate of fuel consumption is relatively low. • Can be situated in sparsely populated areas. • Nuclear power stations are flexible in meeting demand. • No CO_2 or SO_2 produced.	• Although there is very little escape of radioactive material in normal use, radioactive waste can stay dangerously radioactive for thousands of years and safe storage is expensive. • Building and decommissioning is costly. • Longest comparative start-up time.

Summary of Non-renewable Sources of Energy

Advantages	Disadvantages
• Produce huge amounts of energy. • Reliable. • Flexible in meeting demand. • Do not take up much space (relatively).	• Pollute the environment. • Cause global warming and acid rain. • Will eventually run out. • Fuels often have to be transported over long distances.

Renewable Energy Sources

Renewable energy sources are those that will not run out because they are continually being replaced. Many of them are caused by the Sun or Moon. The gravitational pull of the Moon creates tides. The Sun causes:

- evaporation, which results in rain and flowing water
- convection currents, which result in winds, which in turn create waves.

Generating Electricity from Renewable Energy Sources

Renewable energy sources can be used to drive turbines or generators directly. In other words, no fuel needs to be burnt to produce heat.

The table below shows the most common methods of generating energy from renewable energy sources.

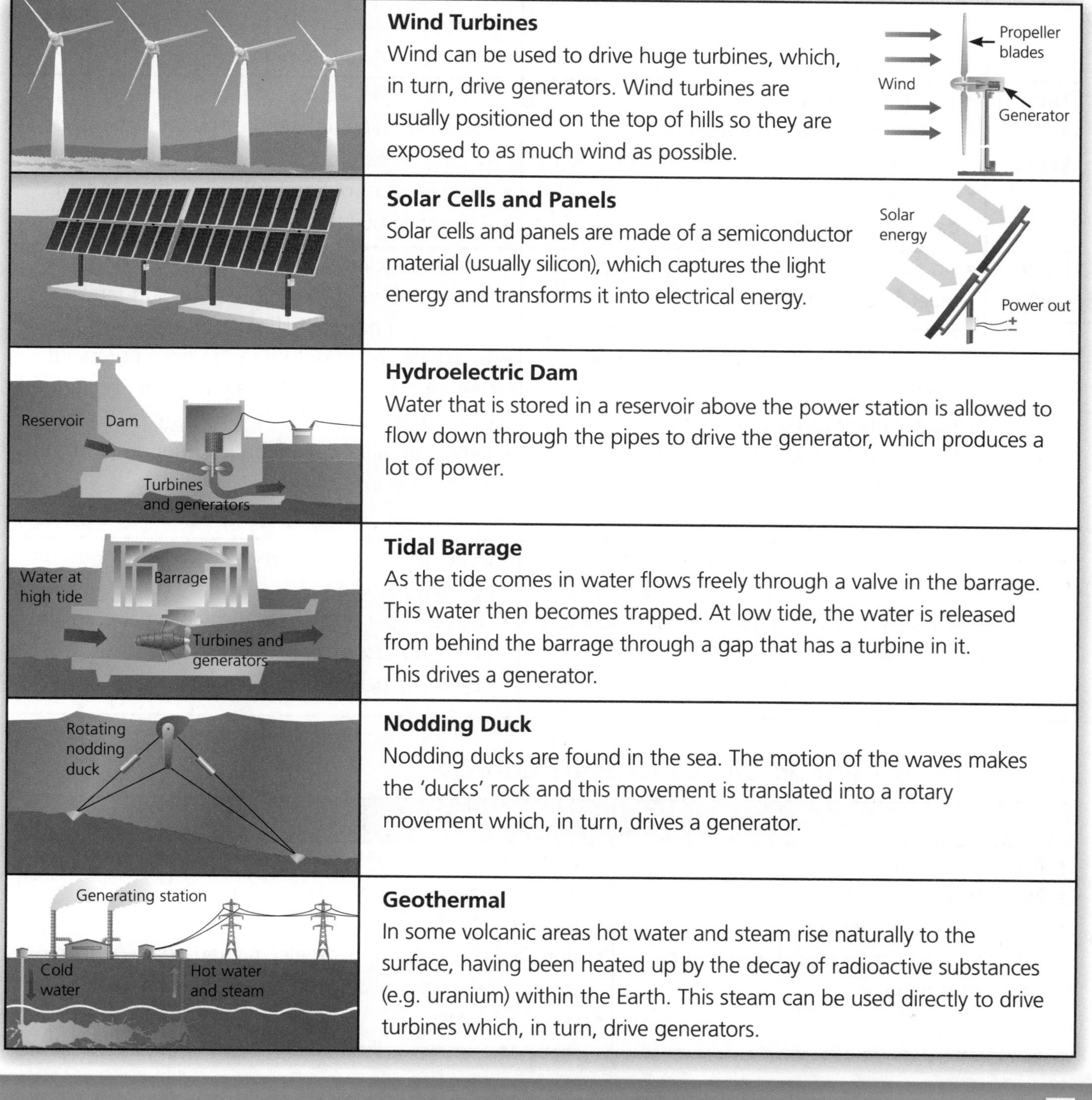

Wind Turbines
Wind can be used to drive huge turbines, which, in turn, drive generators. Wind turbines are usually positioned on the top of hills so they are exposed to as much wind as possible.

Solar Cells and Panels
Solar cells and panels are made of a semiconductor material (usually silicon), which captures the light energy and transforms it into electrical energy.

Hydroelectric Dam
Water that is stored in a reservoir above the power station is allowed to flow down through the pipes to drive the generator, which produces a lot of power.

Tidal Barrage
As the tide comes in water flows freely through a valve in the barrage. This water then becomes trapped. At low tide, the water is released from behind the barrage through a gap that has a turbine in it. This drives a generator.

Nodding Duck
Nodding ducks are found in the sea. The motion of the waves makes the 'ducks' rock and this movement is translated into a rotary movement which, in turn, drives a generator.

Geothermal
In some volcanic areas hot water and steam rise naturally to the surface, having been heated up by the decay of radioactive substances (e.g. uranium) within the Earth. This steam can be used directly to drive turbines which, in turn, drive generators.

Comparing Renewable Sources of Energy

The energy sources listed below use modern technology to provide us with a clean, safe alternative source of energy. Some of their advantages and disadvantages are given.

Source	Advantages	Disadvantages
Wind	• No fuel and little maintenance required. • No pollutant gases produced. • Once built they provide 'free' energy when the wind is blowing. • Can be built offshore.	• Need many turbines to produce a sizable amount of electricity, which means noise and visual pollution. • Electricity output depends on the strength of the wind. • Not very flexible in meeting demand unless the energy is stored. • Building cost can be high.
Tidal and Wave	• No fuel required. • No pollutant gases produced. • Once built they provide 'free', reliable energy. • Barrage water can be released when demand for electricity is high.	• Tidal barrages, across estuaries, are unsightly, a hazard to shipping and destroy the habitats of wading birds, etc. • Daily variations of tides and waves affect output. • High initial building cost.
Hydro-electric	• No fuel required unless storing energy to meet future demand. • Fast start-up time to meet sudden demand. • Produce large amounts of clean, reliable electricity. • No pollutant gases produced. • Water can be pumped back up to the reservoir when demand for electricity is low, e.g. in the night.	• Location is critical and often involves damming upland valleys, which means flooding farms, forests and natural habitats. • To achieve a net output (aside from pumping) there must be adequate rainfall in the region where the reservoir is. • Very high initial capital outlay (though worth the investment in the end).
Solar	• Ideal for producing electricity in remote locations. • Excellent energy source for small amounts. • Produces free, clean electricity. • No pollutant gases produced.	• Dependent on the intensity of light; more useful in sunny places. • High cost per unit of electricity compared to all other sources, except non-rechargeable batteries.

Summary of Renewable Sources of Energy

Advantages	Disadvantages
• No fuel costs during operation. • No chemical pollution. • Often low maintenance. • Do not contribute to global warming or acid rain formation.	• With the exception of hydroelectric and tidal, they produce small amounts of electricity. • Take up lots of space and are unsightly. • Unreliable (apart from hydroelectric and tidal), depend on the weather and cannot guarantee supply on demand. • High initial cost.

New Technologies – Small-Scale Production

As renewable technologies improve, small-scale production is becoming more cost effective. Solar panels are used for road signs and lights in remote areas that are not connected to the National Grid.

Solar panels, wind turbines, ground pump heating and even micro hydroelectric systems are now available for private homes. Installing these systems can save a home owner hundreds of pounds a year in electricity and heating bills. The payback time is 10–25 years depending on the method used.

Carbon Capture

Carbon capture and storage is a rapidly growing technology being used to deal with the carbon dioxide produced by burning fossil fuels. Some of the best storage containers are oil and gas fields.

Carbon dioxide can be pumped into gas fields, displacing the natural gas that is burned as fuel. Carbon dioxide can also be pumped into oil fields, which keeps the oil under pressure making it easier to extract.

Solar Furnaces

Large-scale solar production uses hundreds of mirrors to reflect infrared energy from the Sun on to a tank full of water.

This energy boils the water, which produces steam that is used for driving turbines in the same way as in a normal power station. An advantage is that no polluting gases are released.

By 2013 the Solucar complex in southern Spain will be producing enough energy for 240 000 homes.

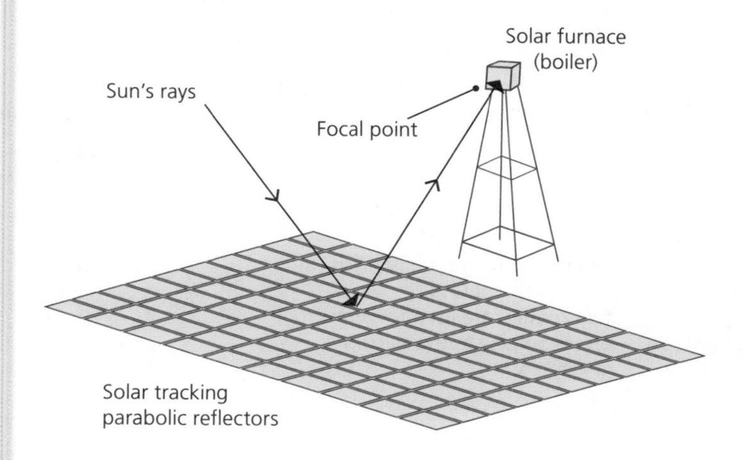

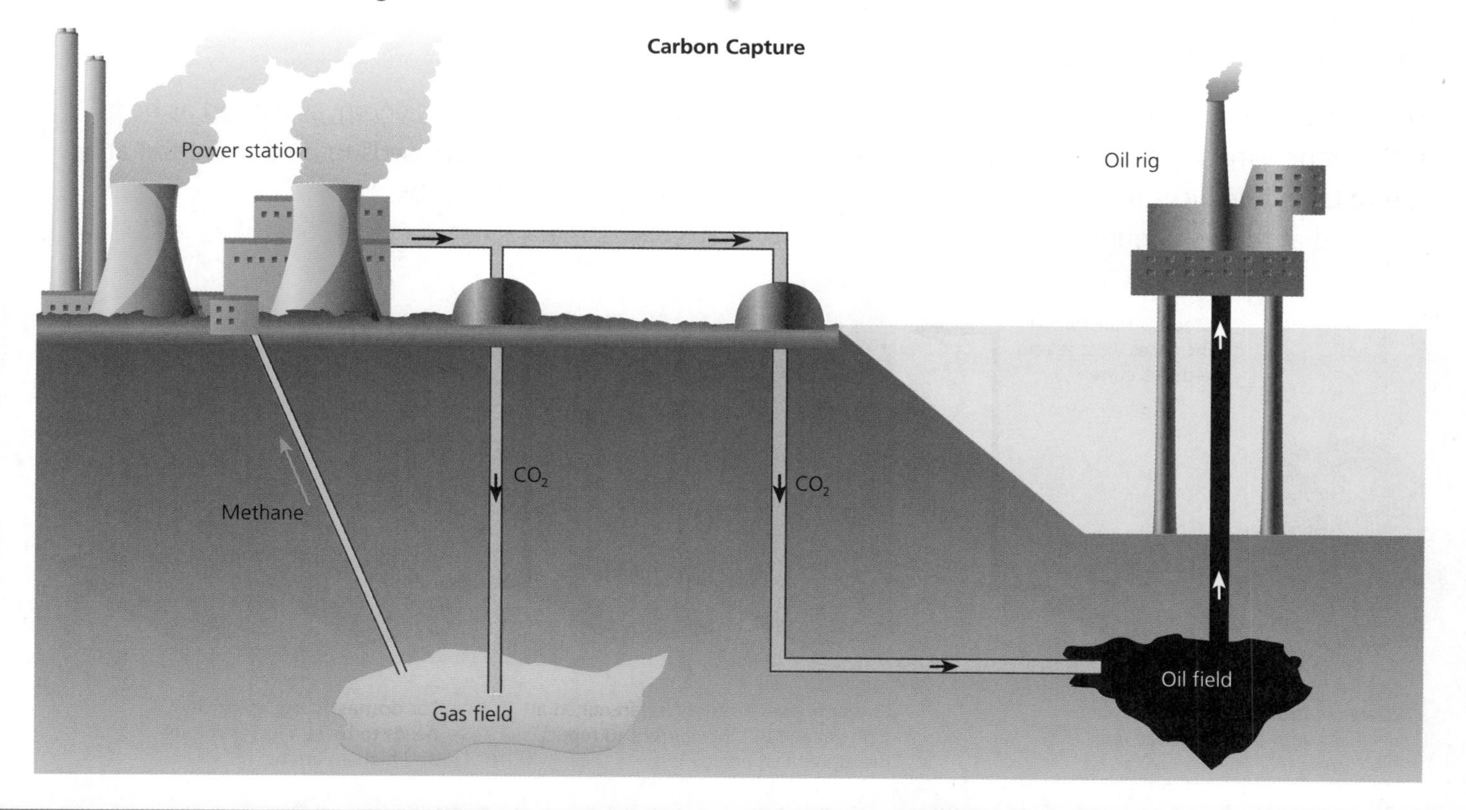

The National Grid

Electricity generated at power stations is distributed to homes, schools and factories all over the country by a network of cables called the **National Grid**. **Transformers** are used to change the **voltage** of the alternating current supply before and after it is **transmitted** through the National Grid. Where it is economically viable to do so, small-scale production systems also provide electricity to the grid.

Overhead power lines are used in the countryside – they are cheaper to install and can cross roads and rivers easily. **Underground** cables are used in towns because of the tall buildings and for safety (people can't access them).

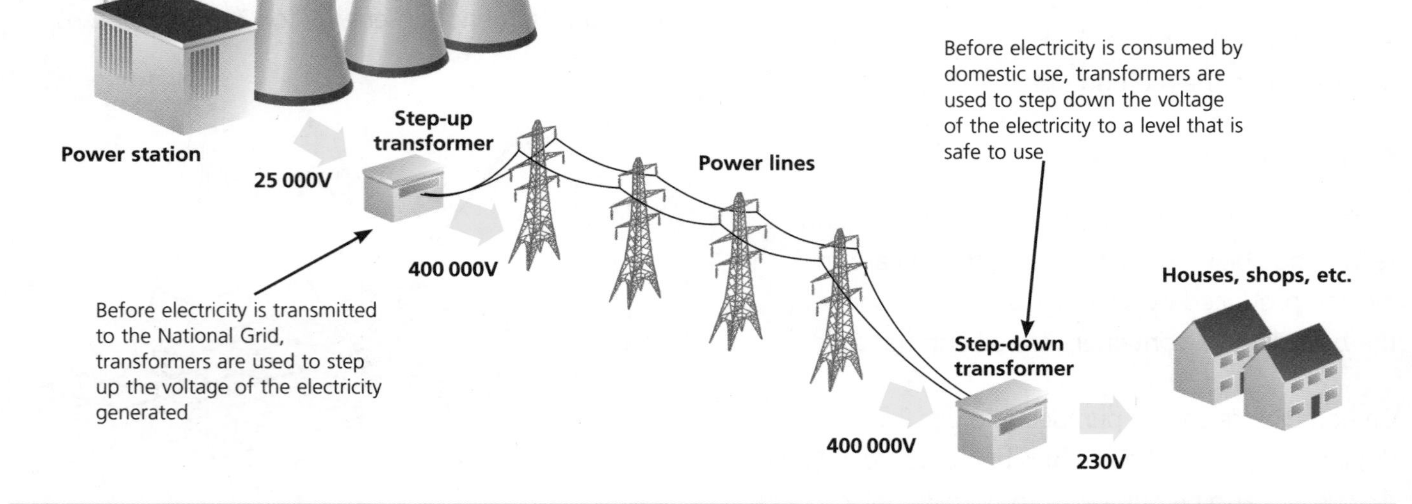

Reducing Energy Loss During Transmission

For a given power, increasing the voltage reduces the current. When a step-up transformer increases the voltage, it reduces the current. The lower the current that passes through a wire the cooler the wire stays and the less energy is wasted as heat. Therefore, electricity needs to be transmitted at as low a current as possible.

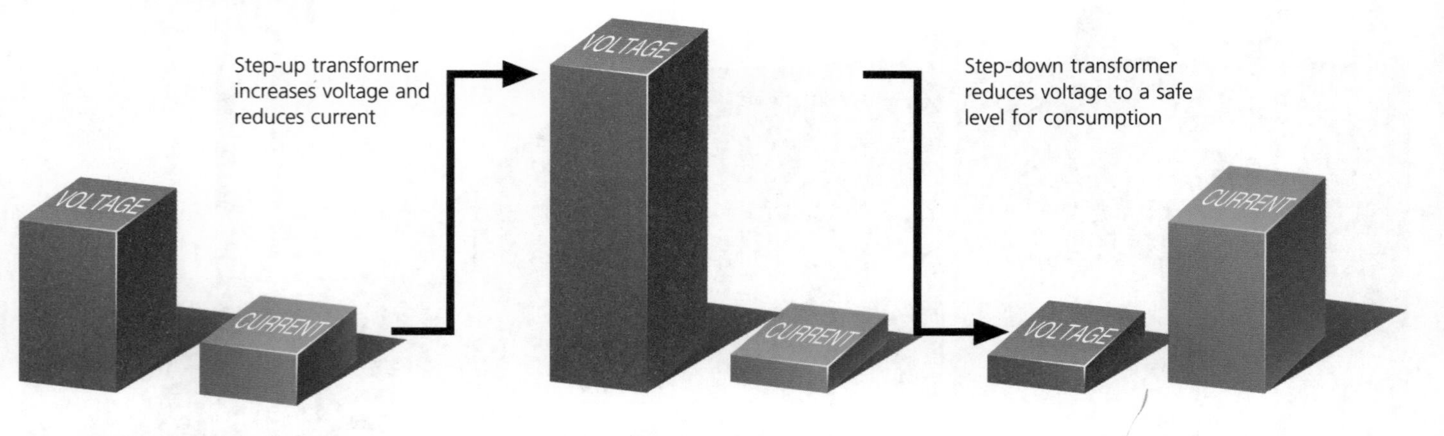

You need to be able to compare and contrast the particular advantages and disadvantages of using different energy sources to generate electricity and consider what would happen if electricity was not available.

Example

The Lonsdale News, Tuesday September 20 2011

Power to the People

The little village of Bukestead was in uproar yesterday after proposed plans were unveiled for a nuclear power station to be built just 10 miles away from the village on the top of the moor. The local area is well known for its beautiful countryside and is a carefully controlled breeding area for many birds. However, increasing demand for electricity means that new electricity generating plants need to be built and this site is considered to be most suitable. Its isolated location and the large number of jobs it will create, have made nuclear power a popular choice with some people.

However, opinion is strongly divided as to the best use of the land, with renewable energy options being cited by many as a better alternative.

Join the debate. Write to your local MP or councillor and have your say.

6 Moorland Road
Bukestead
Derbyshire

Mrs J Smith (MP)
Derby Council Offices
134 High Street
Derby

19th June

Dear Mrs Smith

I was appalled to read about the proposed nuclear power station. The blight on the countryside, the traffic and the horror of a nuclear leak would be unthinkable! What would happen to the local countryside, farms, and the breeding grounds for migrating birds that have been so carefully built up? The local roads would be constantly clogged up with tankers travelling to and from the plant, some carrying highly dangerous waste for disposal. Not to mention the expense of building the site!

This land would be ideal for a farm of wind turbines. The constant strong winds on the hill, coupled with adequate space for 10 to 15 turbines, would be sufficient to produce a sizable amount of electricity. There would be no pollution, no fuel and little maintenance required, just clean, free energy. I hope you will consider the alternatives.

Yours sincerely,

Anne Pentwhistle

(A very concerned local resident)

26 Hill View, Bukestead, Derbyshire

Mrs J Smith (MP)
Derby Council Offices
134 High Street
Derby

17th June

Dear Mrs Smith

I am writing in support of the proposed nuclear power station.

The isolated position on the hill would be ideal. Nuclear leaks are so uncommon that no real threat would be posed and the local economy would greatly benefit from the number of jobs created.

I know that reasons are being cited for the use of 'green' electricity – big solar panels, huge wind turbine farms and the like. These are unfeasible solutions; they could never fulfil the demand, and they themselves would visually pollute the countryside. Everyone says, 'not on my door step', but nuclear is flexible, non-pollutant and will provide the electricity needed for the area.

The need for reliable electricity is very important. Without it we could have endless power cuts, creating the need to use candles for lighting, and leaving some people without heating or cooking facilities. Even worse it could cause accidents when road lights stop working and cause problems in hospitals.

I whole-heartedly support your plans.

Yours sincerely,

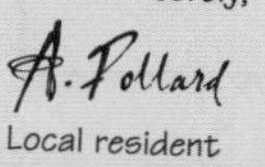

Local resident

P1.5 The use of waves for communication and to provide evidence that the Universe is expanding

All waves transfer energy without transferring matter. Waves can be either transverse or longitudinal. To understand this, you need to know:

- about reflection, refraction and diffraction
- the different types of waves
- about speed, frequency, wavelength and amplitude
- the wave equation.

Wavelength, Frequency and Amplitude

The diagram shows a wave. The **wavelength** is the length of one complete wave, measured in metres (m). The **amplitude** is the height from the middle of the wave to the top.

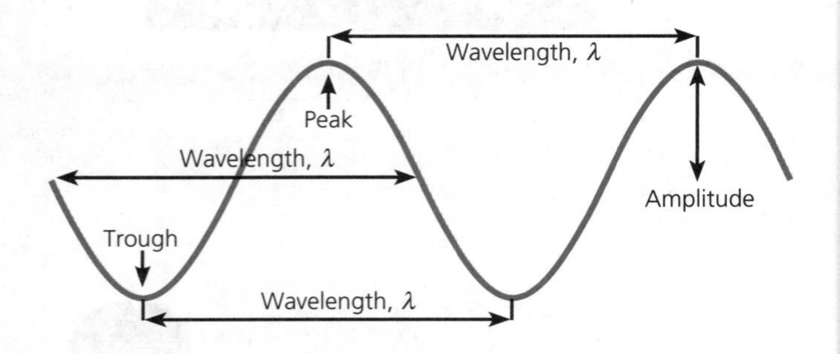

There are two other properties of waves not shown on the diagram:

- **Wave speed** – how quickly the energy moves, measured in metres per second (m/s).
- **Frequency** – how many waves pass a fixed point in a second, measured in hertz (Hz).

All waves obey the wave equation:

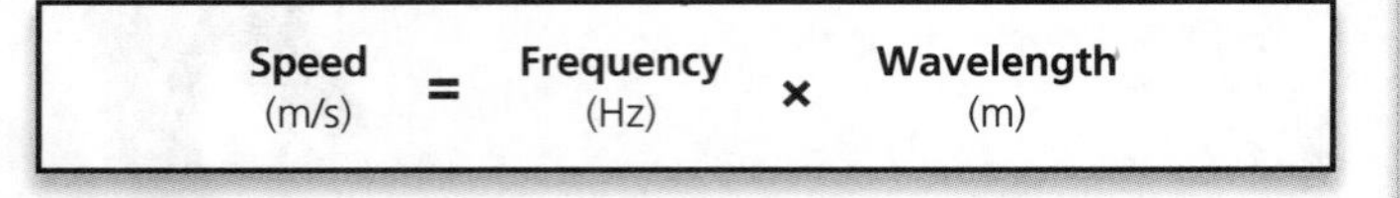

Transverse Waves

Waves in water are an example of a transverse wave. The water moves up and down as the wave travels along.

We say that the oscillations are perpendicular to the direction of energy transfer.

A slinky spring can be used to show transverse waves.

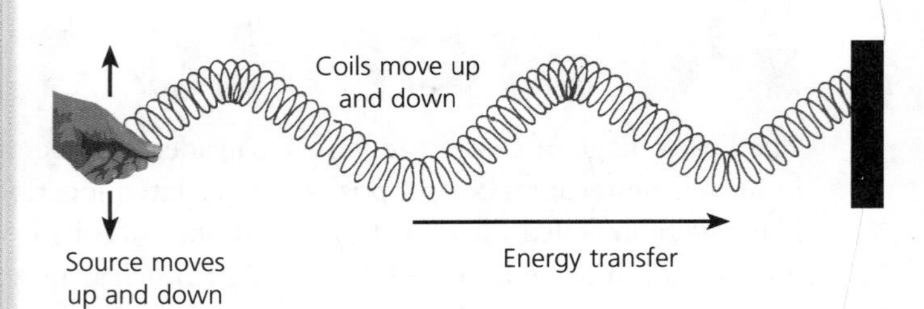

Longitudinal Waves

A slinky spring can also show longitudinal waves. In a longitudinal wave the oscillations are parallel to the direction of energy transfer.

Places where the particles are bunched together are called **compressions**; where they are spread apart they are called **rarefactions**.

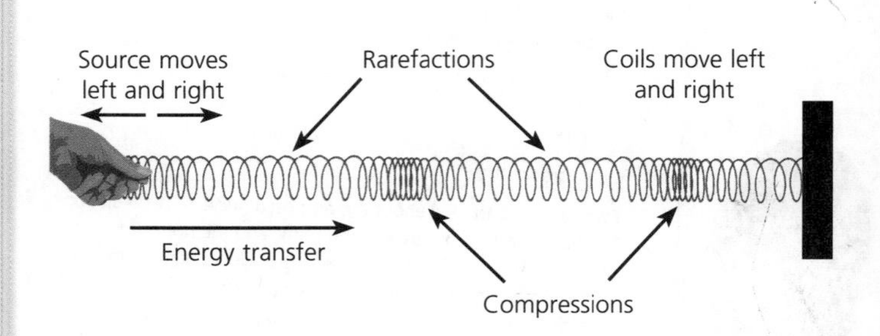

Reflection

When waves are incident on a surface they can be reflected, transmitted or absorbed. An echo is heard when sound is reflected. When light is reflected an image can be formed.

Reflection of Light

When light rebounds off a surface and changes direction it is **reflected**. The diagram shows light being reflected in a plane mirror (flat, smooth, shiny).

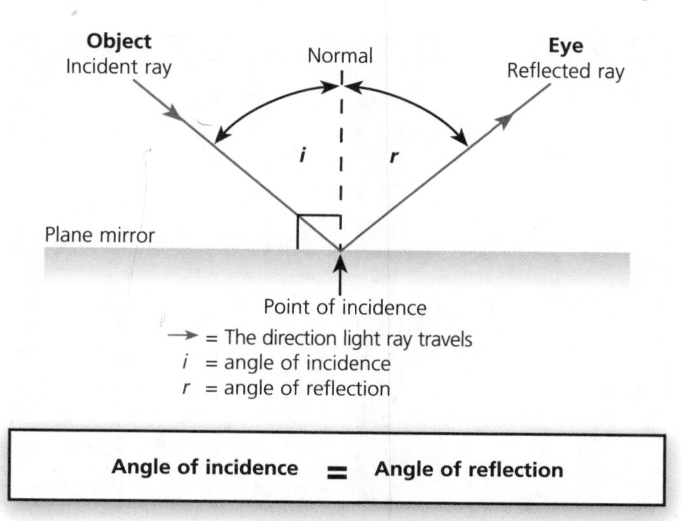

Angle of incidence = Angle of reflection

The **normal line** is perpendicular to the reflecting surface at the point of incidence. The normal is used to calculate the angles of incidence and reflection. The **incident ray** is the light ray travelling towards the mirror. The **reflected ray** is the light ray travelling away from the mirror.

Images Produced by Mirrors

The image formed by a **plane** mirror is the same size as the object. It is upright and laterally inverted (faces the opposite way to the object). The rays of light reaching the eye appear to come directly from the image. Because the rays of light do not actually come from the image, it is described as a **virtual** image.

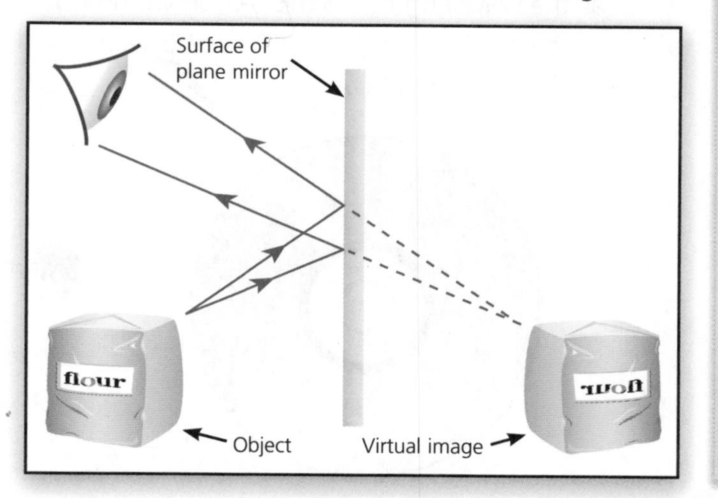

Ray Diagrams

You need to be able to draw ray diagrams illustrating images formed by a plane mirror. By following these steps you should be able to do so accurately. The green arrow shows the object.

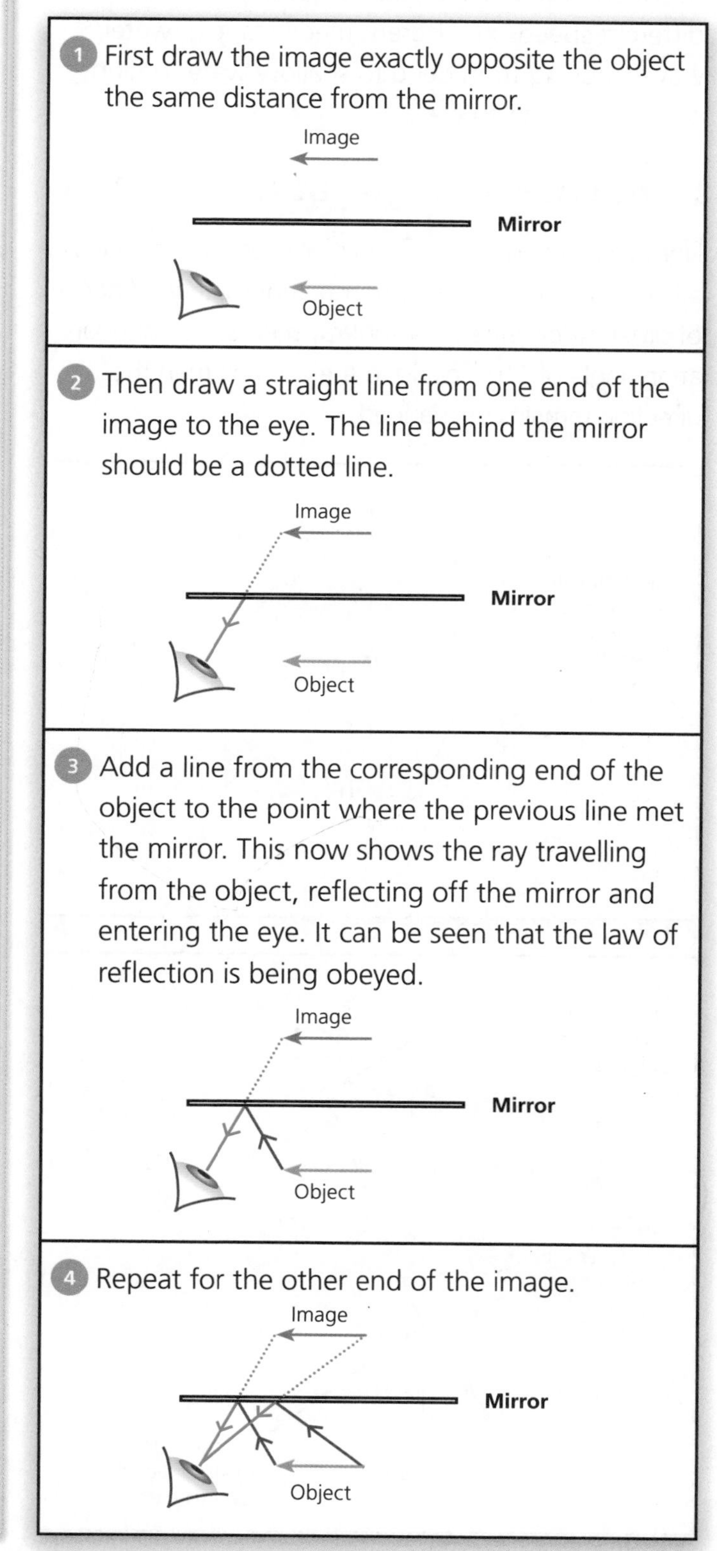

Refraction and Diffraction

As well as being reflected, waves can be **refracted** and **diffracted**.

Refraction occurs when a wave crosses an interface between different substances. Waves travel at different speeds in different mediums, e.g. water waves moving from deep to shallow water or light passing from air to glass.

Refraction of Light at an Interface

Light changes direction when it crosses an interface, i.e. a boundary between two transparent materials (media) of different densities. If a light ray meets the boundary at an angle of 90° (i.e. along the normal) then the direction remains unchanged.

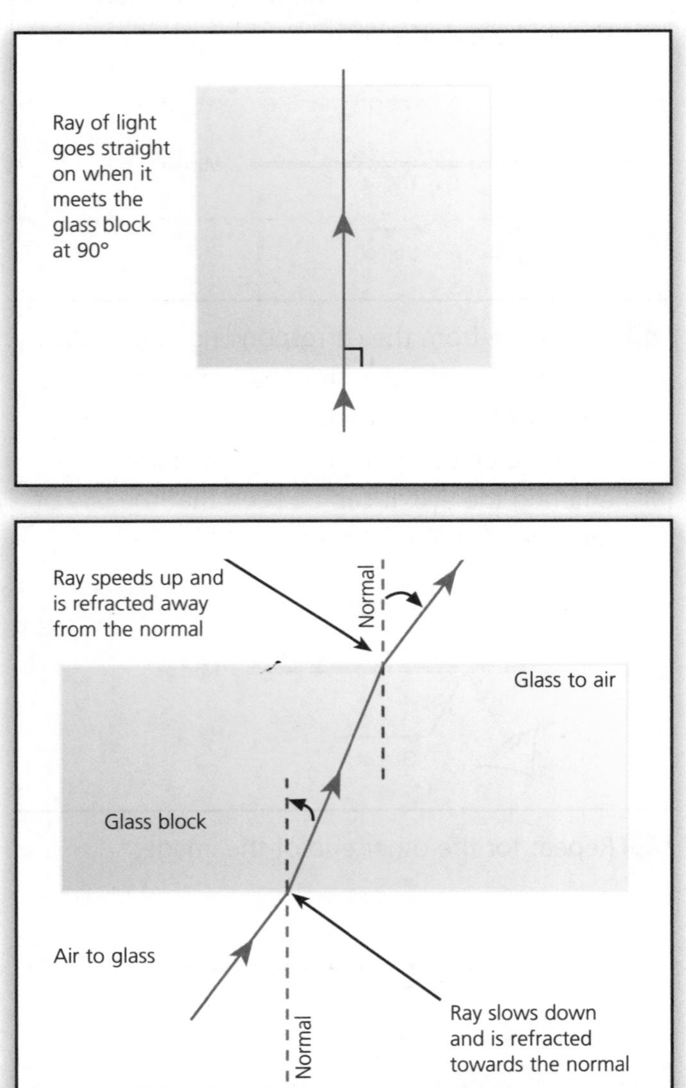

Diffraction

Diffraction occurs when a wave passes through a narrow opening. This can be seen when waves enter a harbour.

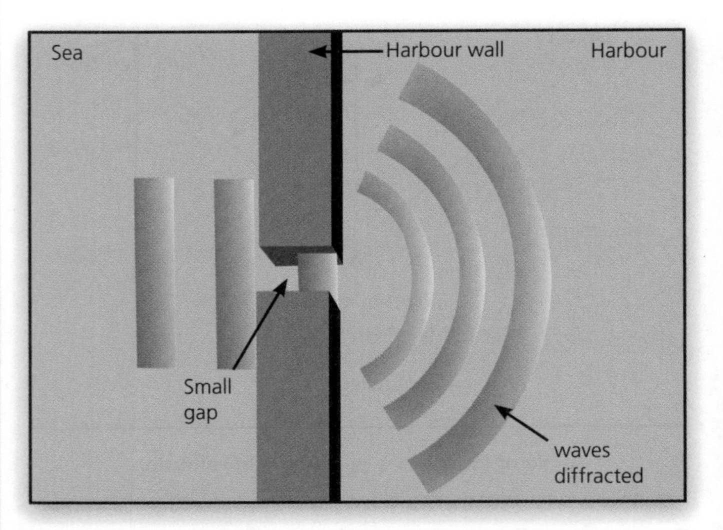

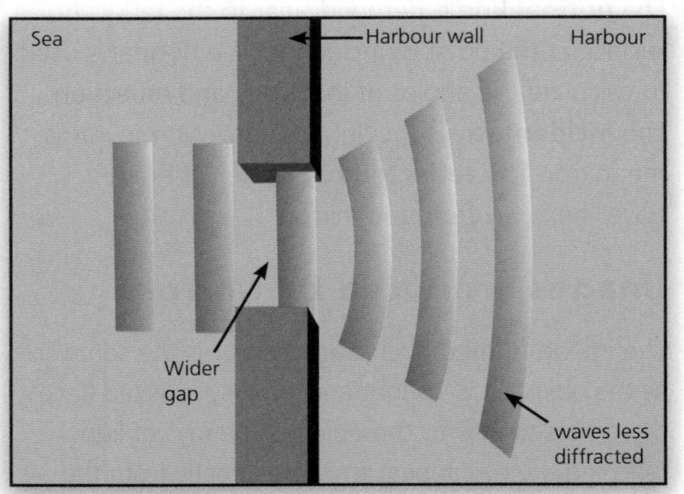

Diffraction is greatest when the gap is the same width as the wavelength. For this reason we do not often see diffraction of light, but it can be seen producing colours on a CD.

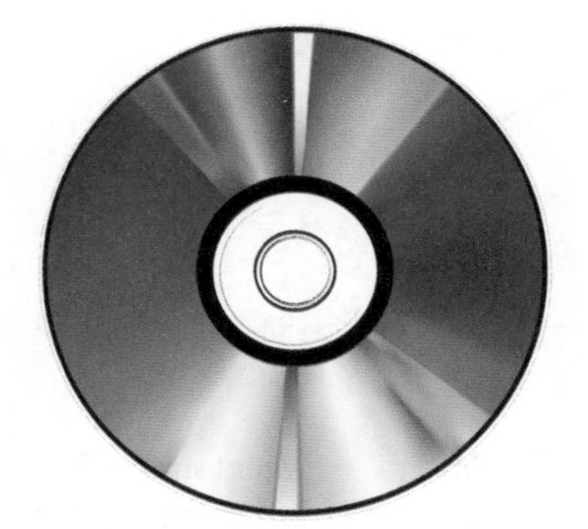

Electromagnetic Waves

Electromagnetic waves are transverse waves – they are the only type of wave that can travel through a vacuum.

Each type of electromagnetic radiation:

- has a different wavelength and a different frequency (the higher the frequency the greater the energy)
- travels at the same speed (300 000 000m/s) through a vacuum (e.g. space).

Electromagnetic waves (such as light) form a continuous range called the **electromagnetic spectrum**.

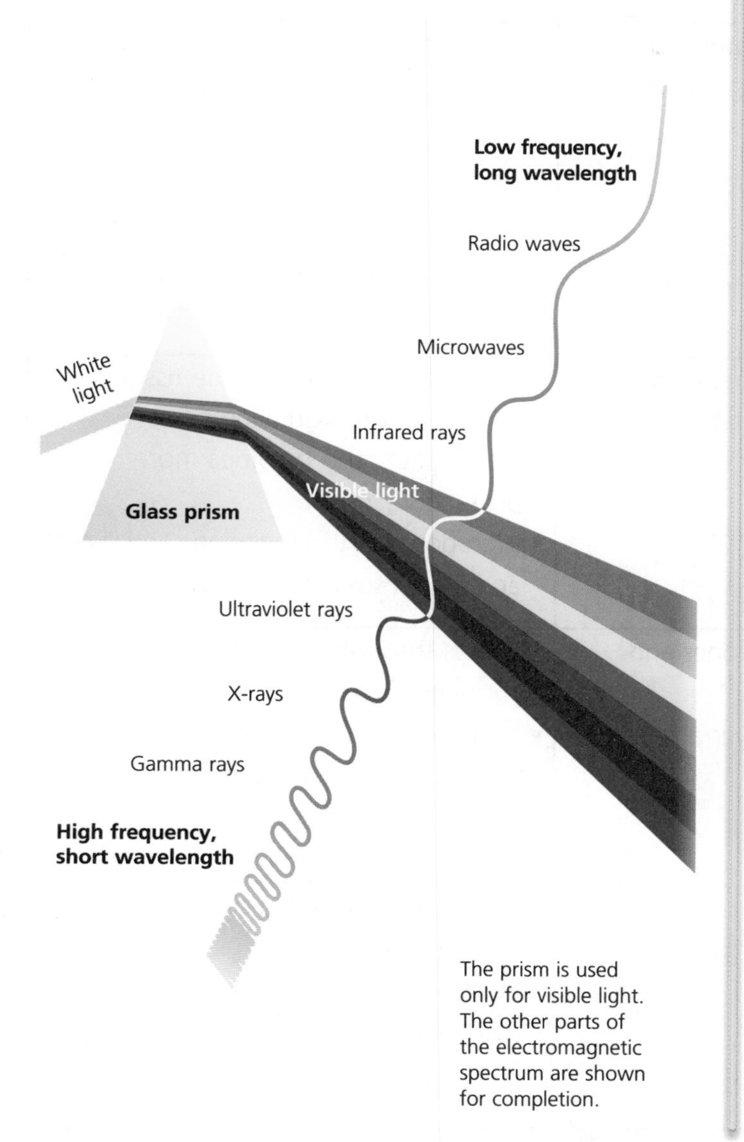

The prism is used only for visible light. The other parts of the electromagnetic spectrum are shown for completion.

Radio waves, microwaves and infrared rays all have a longer wavelength and a lower frequency than visible light.

Ultraviolet waves, X-rays and gamma rays all have a shorter wavelength and a higher frequency than visible light.

Electromagnetic waves can be **reflected** and **refracted**. Different wavelengths of electromagnetic waves are reflected, absorbed or transmitted differently by different substances and types of surface, e.g. black surfaces are particularly good absorbers and emitters of infrared radiation.

When a wave is **absorbed** by a substance, the energy it carries is absorbed and makes the substance heat up. It may also create an **alternating current** of the same frequency as the radiation. This principle is used in television and radio aerials, which receive information via radio waves.

Electromagnetic waves obey the wave formula:

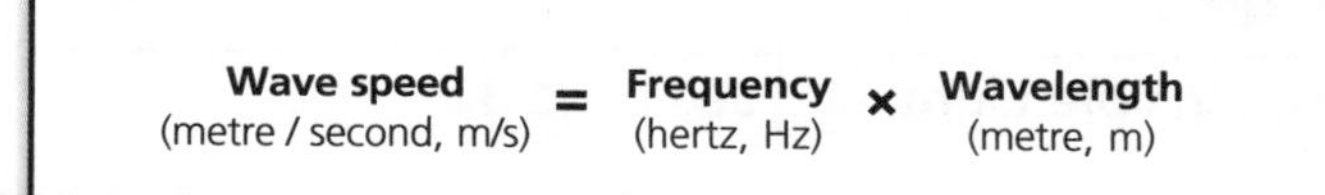

Visible Light

Light is one type of electromagnetic wave that, together with the other various types of radiation, is in the electromagnetic spectrum.

The seven 'colours of the rainbow' form the **visible spectrum** which, as the name suggests, is the only part of the electromagnetic spectrum that we can see.

The visible spectrum is produced because white light is made up of many different colours. These are refracted by different amounts as they pass through a prism – red light is refracted the least and violet is refracted the most.

Visible light can be used for communication through optical fibres, and for vision and photography.

The Use of Waves

Electromagnetic wave	Uses	Effects
Radio Waves	• Transmitting radio and TV signals between places – waves with longer wavelengths are reflected by the ionosphere (an electrically charged layer in the atmosphere) so they can send signals between points regardless of the curve of the Earth's surface.	• High levels of exposure for short periods can increase body temperature leading to tissue damage, especially to the eyes.
Microwaves	• Satellite communication networks and mobile phone networks. • Cooking – microwaves are absorbed by water molecules causing them to heat up. • Bluetooth devices.	• May damage or kill cells because they are absorbed by water in the cells, leading to the release of heat. Therefore, care must be taken in the use of microwaves.
Infrared Rays	• Grills, toasters and radiant heaters. • Remote controls for televisions and video recorders. • Optical fibre communication.	• Absorbed by skin and felt as heat. • Excessive amount can cause burns.
Visible Light	See page 31	See page 31
Ultraviolet Rays	• Security coding – a surface coated with special paint absorbs UV and emits visible light. • Sunbathing and sunbeds.	• Passes through skin to the tissues below. Darker skin allows less penetration and provides more protection. • High doses of this radiation can kill cells and low doses can cause skin cancer.
X-Rays	• Producing shadow pictures of bones and metals. • Some cancers can be treated by irradiation with X-rays (radiotherapy).	• Passes through soft tissues, although some is absorbed. • High doses can kill cells and low doses can cause cancer.
Gamma Rays	• Killing cancerous cells. • Killing bacteria on food and surgical instruments.	• Passes through soft tissues – although some is absorbed. • High doses of this radiation can kill cells and low doses can cause cancer.

You need to be able to evaluate the possible hazards associated with the use of different types of electromagnetic radiation.

Example

It's Good to Talk – or is it?

New findings raise concerns that mobile phones could cause cancer and other health problems.

Swedish scientists studying the effects of electromagnetic radiation on red blood cells have found that levels of radiation, equivalent to those emitted by mobile phones, have a significant effect on the attractive forces between cells.

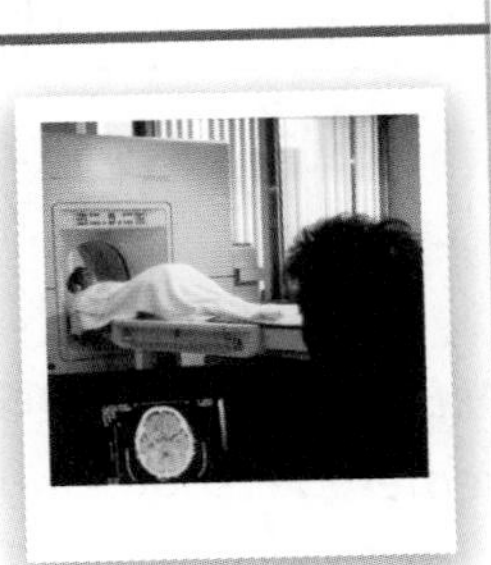

Up until now, the conventional view has been that microwaves could only cause damage at a cellular level if they carried enough energy to break chemical bonds or 'fry' tissue. These new findings might suggest that mobile phones do in fact emit enough energy to affect the bonds.

Experts, however, have been quick to point out that the results were obtained through tests on small groups of cells and provide no real evidence of a danger to health.

There have been other suggestions in the past that mobile phones can cause brain tumours and Alzheimer's disease, but so far research has been inconclusive.

Mobile Phones are Safe

New study has found no link between mobile phones and cancer.

Scientists studying the possible links between mobile phones and brain tumours have reported today that they have found no correlation between using mobile phones and the risk of developing glioma – the most common type of brain tumour.

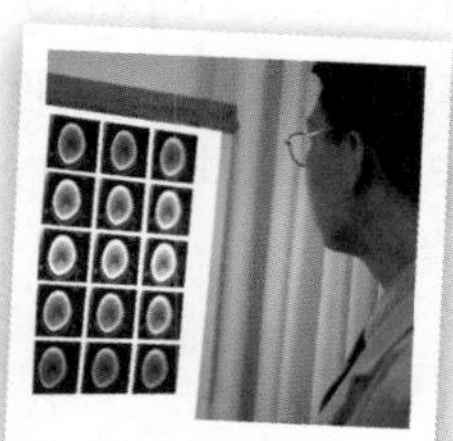

The study of 2682 people across the UK looked at 966 individuals with diagnosed glioma and 1716 individuals without the condition. It concluded that although a large percentage of the cancer sufferers reported their tumours to be on the side of the head where they held the phone, for regular mobile phone users, there was no increased risk of developing glioma.

Using Mobile Phones

Advantages	Disadvantages
• Easy, convenient method of communication, especially in a vulnerable situation – when car breaks down, alone at night, feel threatened, etc. • Can be used to access the Internet, take pictures, and watch television / video clips. • Easy way to keep in contact when away from home or abroad, e.g. text messages. • Many different tariffs and networks, which makes them affordable. • Can help in solving crime because mobile phones can be tracked.	• Some studies have linked mobile phone use with brain tumours and Alzheimer's disease. • The long-term effects of using mobile phones are not known – studies are still being carried out. • Increasingly, advertising is targeted at younger age groups who would be more vulnerable to any health implications.

Sound Waves

Sounds travel as longitudinal waves. They are produced when something vibrates. The quality of a note depends on the waveform.

Sound cannot travel through a vacuum.

- Sound **reflects** off hard surfaces to produce echoes.
- Sound is **refracted** when it passes into a different medium or substance.
- Sound can be **diffracted** around buildings or land masses, so a person in the 'shadow' of a large building can still hear sounds we might expect to be 'blocked'.

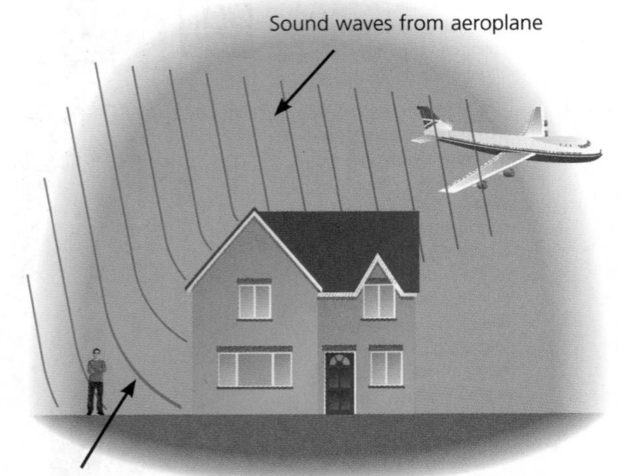

Frequency and Pitch

The frequency of a sound wave is the number of vibrations produced in one second. It is measured in hertz (Hz). Humans can hear sounds in the range of 20–20 000 Hz. The frequency affects the pitch of the sound.

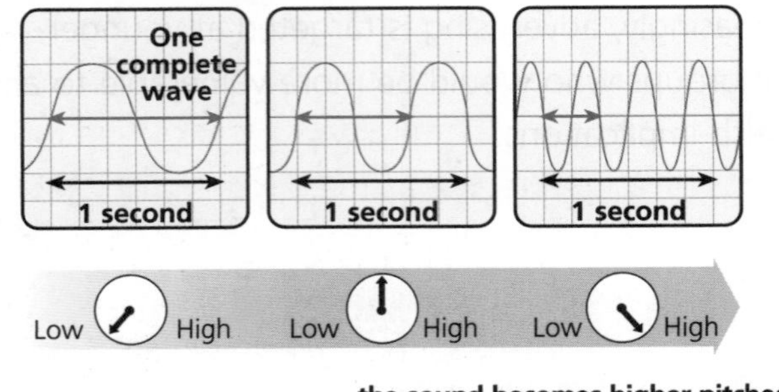

Amplitude and Loudness

Amplitude is the peak of movement of the sound wave from its rest point. The amplitude affects the loudness of the sound.

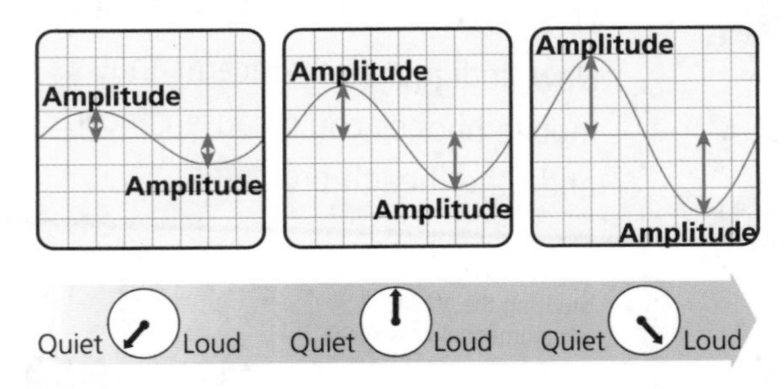

N.B. These diagrams represent the sound wave – remember that sound waves are longitudinal.

The Doppler Effect

When a wave source is moving towards or away from an observer there is a change in the observed wavelength and frequency.

We can detect this change with moving vehicles. For example, when a car approaches you quickly, the sound waves it produces are bunched up, become higher frequency and higher pitched. As the car moves away, the sound waves are stretched out, become lower frequency and lower pitched.

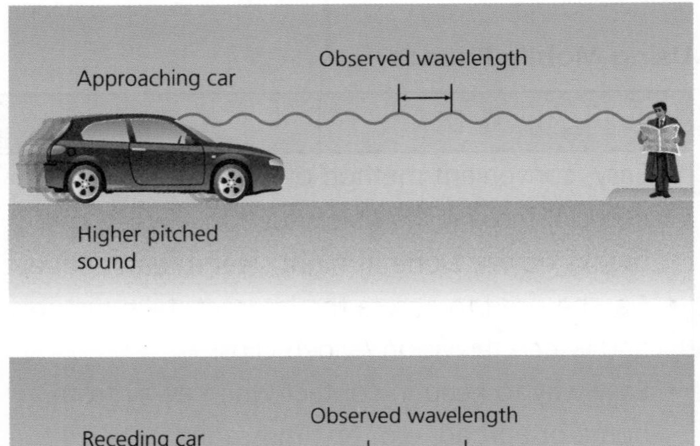

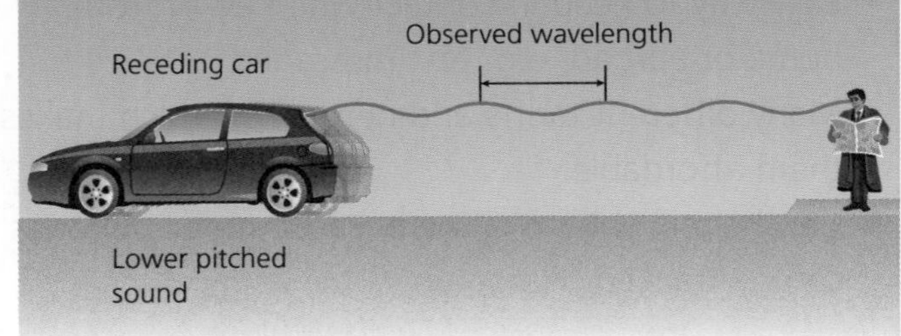

Red-shift

When a light source is moving towards or away from us its frequency appears to change.

When we look at light from the Sun some frequencies of light are not seen because they are absorbed by hydrogen and helium in the Sun. This appears as black absorption lines in the Sun's spectrum, as seen below.

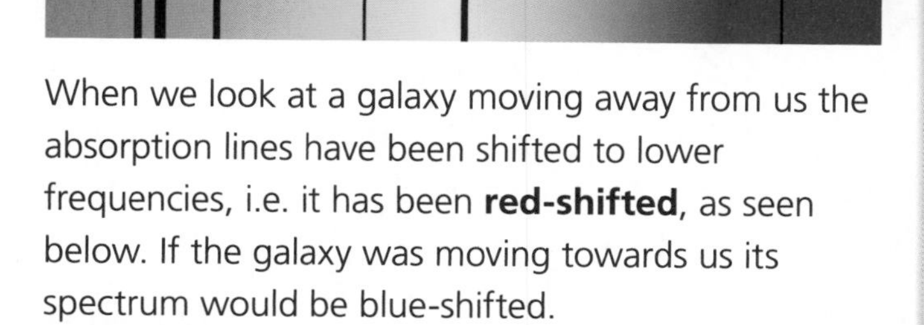

When we look at a galaxy moving away from us the absorption lines have been shifted to lower frequencies, i.e. it has been **red-shifted**, as seen below. If the galaxy was moving towards us its spectrum would be blue-shifted.

Almost all galaxies show red-shift and the further away a galaxy is the more red-shifted it is. This means that the more distant the galaxy the faster it is moving away from us.

The Big Bang Theory

There is evidence that all galaxies are moving away from one another and that the further away they are the faster they are moving. It has therefore been suggested that the entire Universe is expanding. Scientists reason that if the Universe is expanding, then it could have started at a single point and expanded outwards like in an explosion. This conclusion has led to the **Big Bang theory**.

The Big Bang theory states that the Universe started from a very small initial point billions of years ago and has been expanding ever since. The point from which it started therefore contained all the space, energy and matter in the Universe.

For many years scientists could not agree on the Big Bang theory with many preferring alternative theories like the **Steady State theory**.

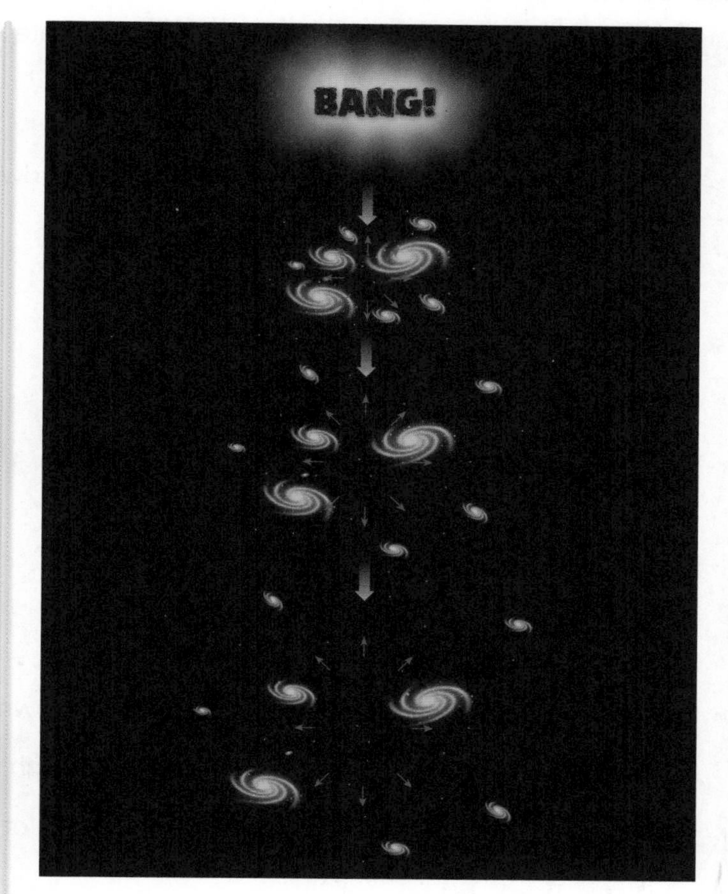

Cosmic Microwave Background Radiation

Cosmic microwave background radiation (CMBR) was discovered in the 1960s. When looking at the space between galaxies with an optical telescope everything appears black. But when using a telescope that is capable of detecting microwaves, a faint glow is seen. This glow is almost the same in all directions and fills the Universe.

It is believed that this background radiation comes from radiation that was present shortly after the beginning of the Universe.

Although there are still some questions that the Big Bang theory cannot answer, it is currently the only theory that can explain microwave background radiation. Therefore, it was the discovery of CMBR that led to the scientific community finally accepting the Big Bang theory as the most probable theory that explains what happened when the Universe began.

P1 Exam Practice Questions

1. A wave machine in a 25 m swimming pool generates waves at a frequency of 2 Hz. During operation there are 5 complete waves in the pool.

   a) Calculate the speed of the water waves.
   Write down the equation you use and then show how you work out your answer. **(4 marks)**

   b) What type of waves are water waves? **(1 mark)**

2. The diagram shows the National Grid system.

   a) On the diagram label the step-up and step-down transformers. **(2 marks)**

   b) When transmitting electricity it is done at a high voltage and low current. Why is this? **(2 marks)**

3. Astronomers have observed that the wavelengths of the light given out from distant galaxies are longer than expected.

   a) What is this called? **(1 mark)**

   b) Circle the correct option in the following sentence. 'This observation gives evidence for the idea that the Universe is … **(1 mark)**

   shrinking **not changing** **expanding**

   c) This observation helped lead some astronomers to the Big Bang theory. What other piece of evidence supports the Big Bang theory? **(1 mark)**

4. The diagram shows the energy transfers in a coal-fired power station.

Energy in fuel
Delivered to customers
Used in the power station
Lost in transmission
Heat losses to the environment

   a) What type of energy is stored in the fuel? **(1 mark)**

   b) What type of energy is delivered to the customers? **(1 mark)**

   c) From the diagram estimate the efficiency of the system and explain how you made your estimate. **(2 marks)**

P2.1 Forces and their effects

Forces can cause changes to the shape or motion of an object. To understand this, you need to know:

- what is meant by resultant force
- how force and acceleration are linked
- what happens when an object falls
- how forces and braking are linked
- what happens when an object is stretched or squashed.

Newton's Third Law

Newton's third law states: 'Whenever two objects interact, the forces they exert on each other are equal and opposite.'

Examples

- **Magnetism** – a metal car is attracted to a magnet; the magnet is attracted to the car with an equal and opposite force.
- **Gravity** – a ball is attracted to a planet by the force of gravity; the planet is attracted to the ball with an equal and opposite force. The ball has a much smaller mass than the planet so it is the ball that moves.

Newton's third law is often confused with the idea of balanced forces. The important thing to remember with these equal and opposite forces is that they are the **same force** acting on **different objects**.

Resultant Forces

A resultant force is the sum of different forces acting on the same object.

When a number of forces are acting on an object it is easier to think of these forces as a single force that has the same effect on the object's motion as all the original forces acting together. This single force is called the **resultant force**.

You need to be able to calculate the size and direction of a resultant force.

Example

Forces acting in the same direction as each other are added up and if they are in the opposite direction they are subtracted.

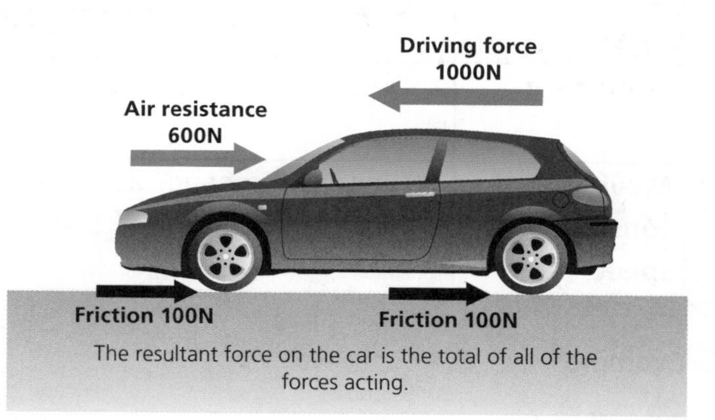

The resultant force on the car is the total of all of the forces acting.

In the diagram we have 1000N acting to the left and 600N + 100N + 100N acting to the right. This gives a total of 1000N to the left and 800N to the right. The overall resultant force is, therefore, 200N acting to the left (i.e. 1000N – 800N).

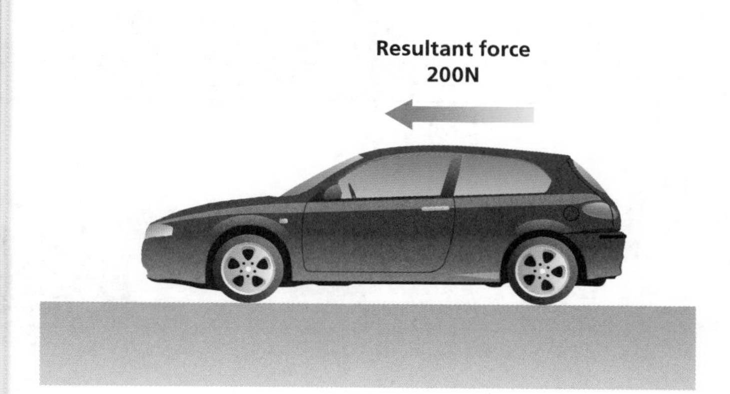

When the resultant force is **zero** we say that the forces are **balanced**. If the resultant force is **not zero** then the forces are **unbalanced**.

Newton's first law describes how a resultant force will affect the movement of an object. It states: 'An object will remain in the same state of motion unless an unbalanced force acts on it'. In short this means that when the resultant force is zero:

- if the object is not moving it stays where it is
- if the object is moving it keeps moving at the same speed in the same direction.

How Forces Affect Movement

The table and diagrams illustrate what happens when the resultant force is not zero.

Object	Resultant Forces	
	Zero (Balanced)	**Not Zero (Unbalanced)**
Stationary	• Object remains stationary	• Object will accelerate in the direction of the resultant force
Moving at a constant speed	• Object will continue at the same constant speed in the same direction	• Object will accelerate in the direction of the resultant force, making it slow down, speed up or change direction

Example

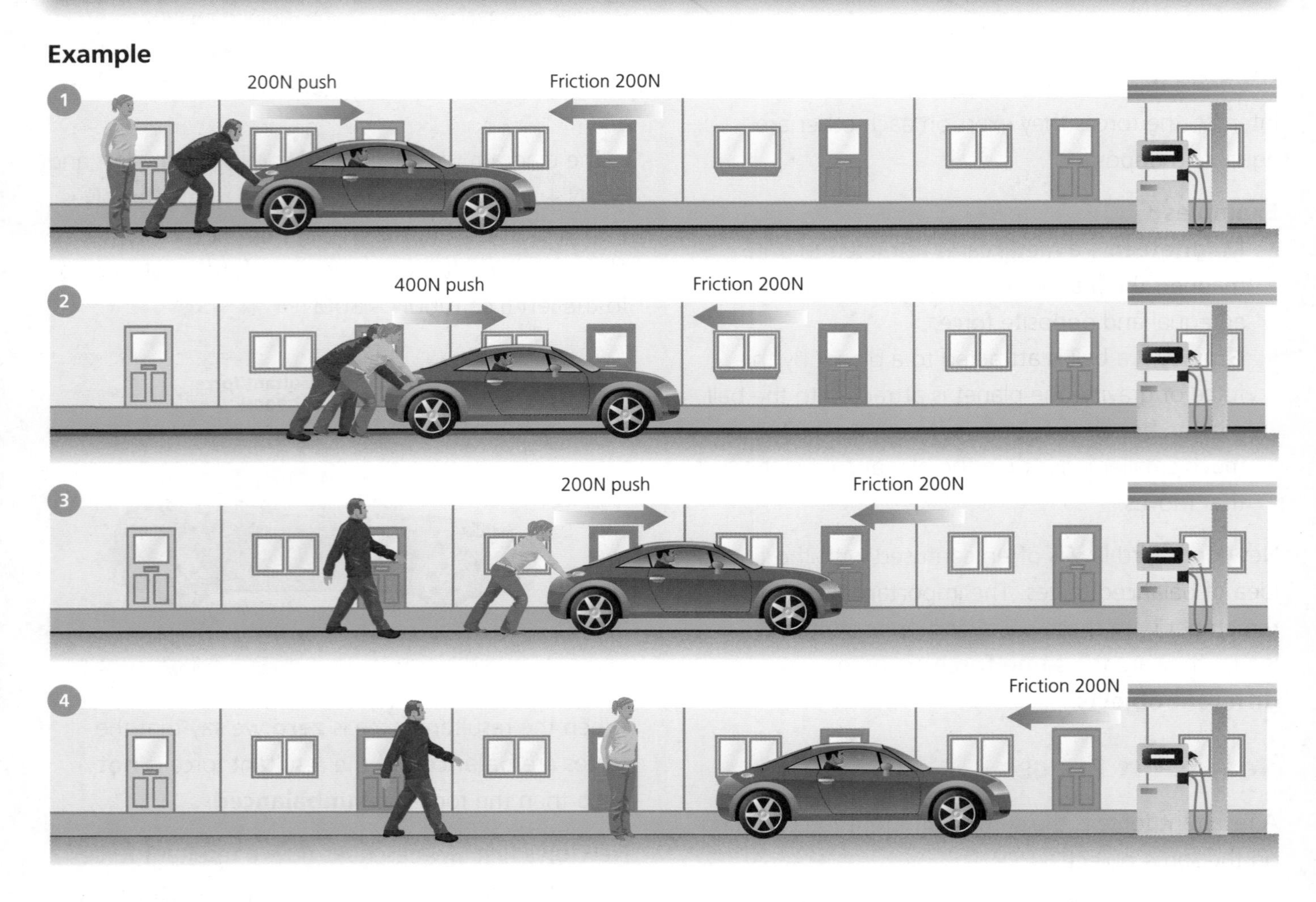

1 With just one person pushing, the push force is equal to the frictional force, and so the forces are balanced and the stationary car does not move.

2 With two people now pushing, the push force is greater than the frictional force, and so an unbalanced force acts and the stationary car will now start to move and speed up (accelerate).

3 When one person drops out, the push force is equal to the frictional force again, but these balanced forces will keep the moving car travelling at the same constant speed.

4 With the petrol pump getting nearer, the second person drops out, making the push force less than the frictional force, and an unbalanced force acts. This causes the car to slow down (decelerate) and eventually stop.

Force, Mass and Acceleration

Newton's second law describes the link between force, mass and acceleration. It states: 'The acceleration of an object is directly proportional to the force and inversely proportional to the mass'.

If an unbalanced force acts on an object then the acceleration of the object will depend on:

- the **size** of the unbalanced force – the bigger the force, the greater the acceleration
- the **mass** of the object – the bigger the mass, the smaller the acceleration.

Example

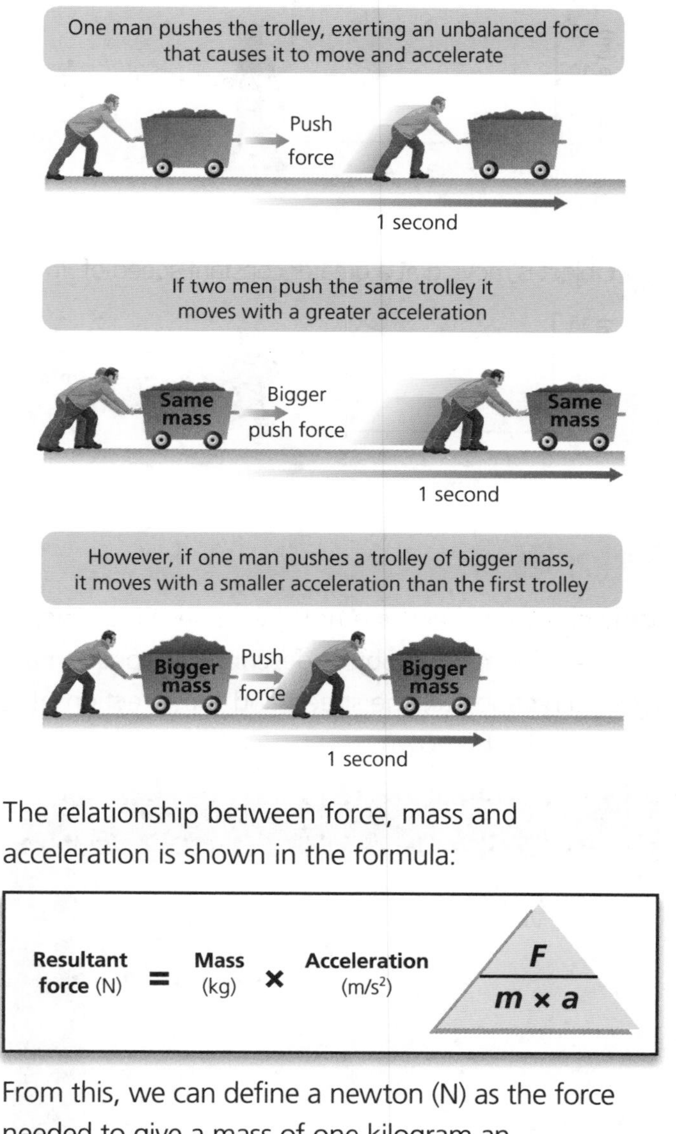

The relationship between force, mass and acceleration is shown in the formula:

Resultant force (N) = **Mass** (kg) × **Acceleration** (m/s^2)

$$F = m \times a$$

From this, we can define a newton (N) as the force needed to give a mass of one kilogram an acceleration of one metre per second squared ($1m/s^2$).

Example

A trolley of mass 400kg is pushed along a floor at a constant speed by one man who exerts a push force of 150N.

Another man joins him to increase the push force and the trolley accelerates at $0.5m/s^2$. Calculate:

a) the force needed to achieve this acceleration

b) the total push force exerted on the trolley.

Initially the trolley is moving at a constant speed, so the forces acting on it must be balanced.

Therefore, the 150N push force must be opposed by an equal force, i.e. friction or air resistance.

When the trolley starts accelerating the push force must be greater than friction, etc. These forces do not cancel each other out and an unbalanced force now acts.

a) Using the formula:

Force = Mass × Acceleration

= $400kg \times 0.5m/s^2$

= 200N

b) Total push = Force needed to equal friction + Force needed to provide acceleration

= 150N + 200N

= 350N

Speed

One way of describing the movement of an object is by measuring its **speed**, i.e. how fast it is moving. The cyclist shown below travels a distance of 8 metres every 1 second, so we can say that the speed of the cyclist is 8 metres per second (m/s).

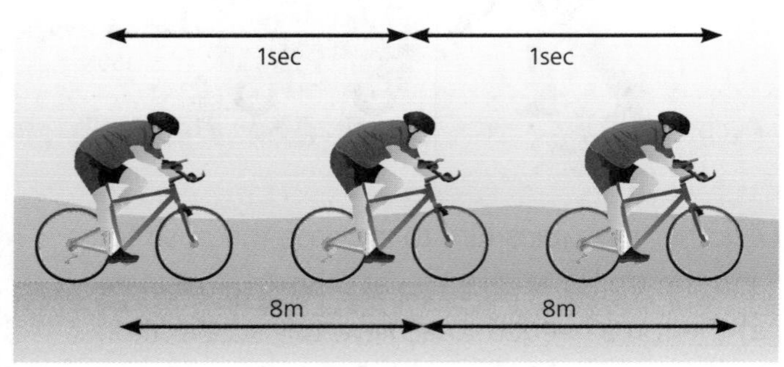

To work out the speed of any moving object you need to know two things:

1. The distance it travels.
2. The time it takes to travel that distance.

The speed of the object can then be calculated using the formula:

$$\text{Speed (m/s)} = \frac{\text{Distance travelled (m)}}{\text{Time taken (s)}}$$

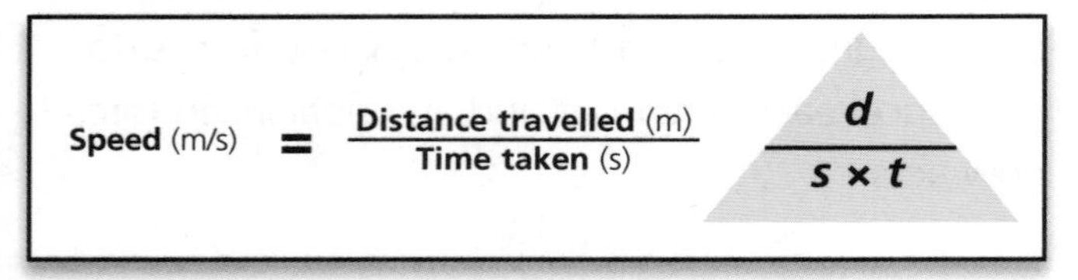

Speed can also be measured in kilometres per hour (km/h).

Example

Calculate the speed of a cyclist who travels 2400m in 5 minutes.

$$\text{Speed (m/s)} = \frac{\text{Distance travelled (m)}}{\text{Time taken (s)}} = \frac{2400}{300} = \mathbf{8m/s}$$

Multiply 5 minutes by 60 to get time in seconds

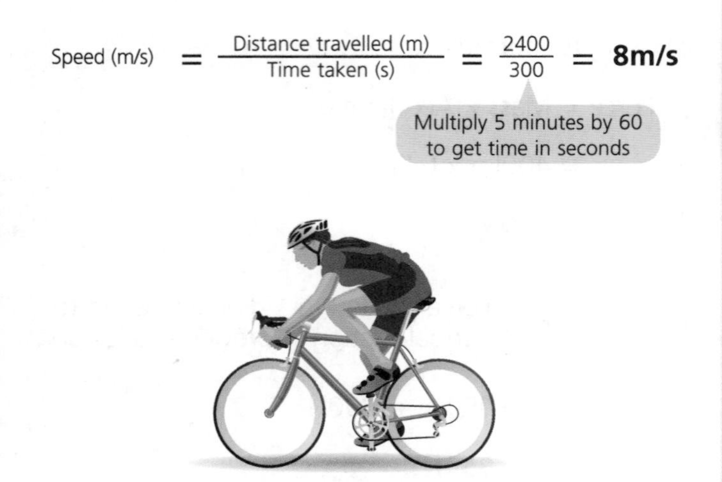

Distance–Time Graphs

The gradient of a **distance–time graph** represents the speed of an object. The steeper the gradient of a graph, the greater the speed of the object.

1. Stationary object. N.B the *'y'* axis shows distance from a fixed point (0,0), not total distance travelled.

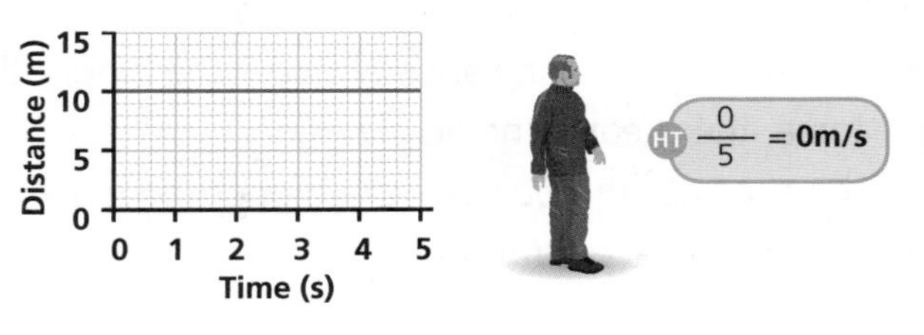

2. Object is moving at a constant speed of 2m/s

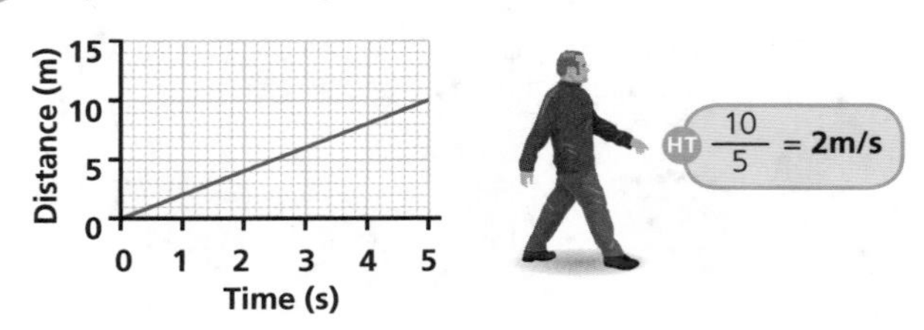

3. Object is moving at a greater constant speed of 3m/s

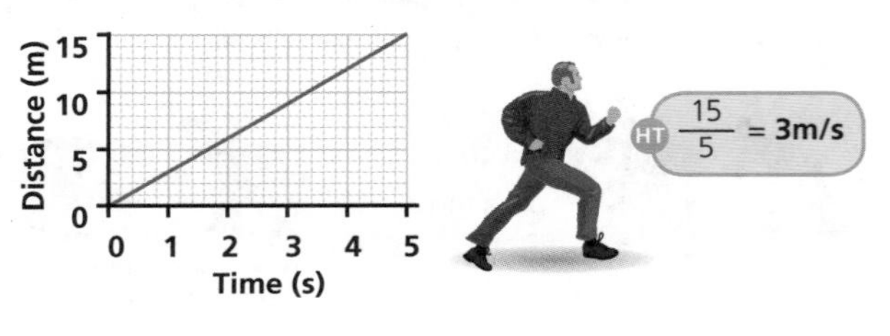

Velocity

Velocity and speed are not the same thing. The velocity of a moving object describes its speed in a given direction, i.e. the speed and the direction of travel are both known.

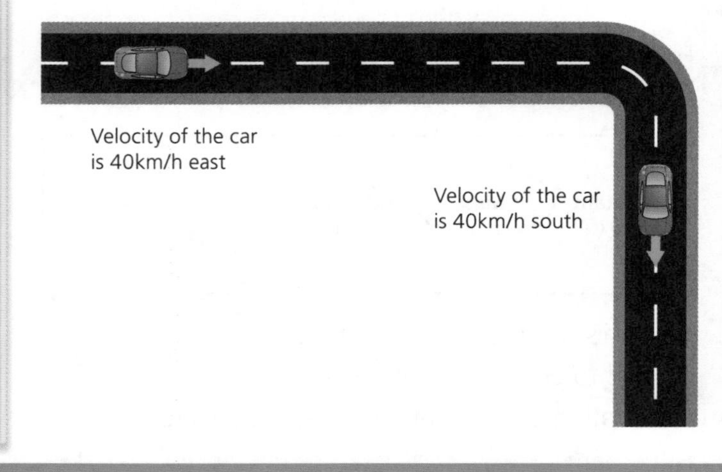

Acceleration

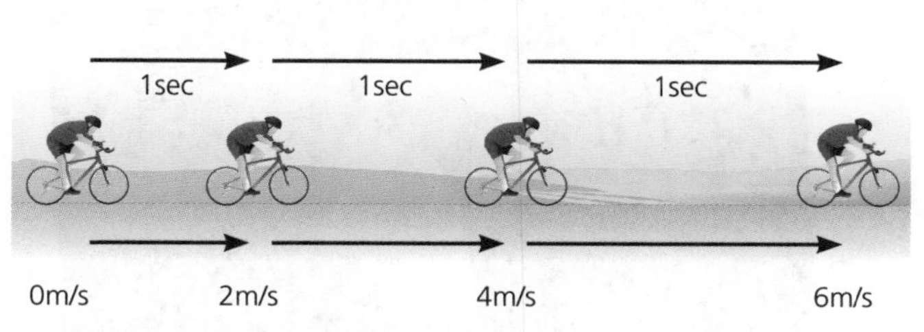

The **acceleration** of an object is the rate at which its velocity changes. In other words it is a measure of how quickly an object speeds up or slows down.

The cyclist above increases his velocity by 2 metres per second every second. So, we can say that his acceleration is 2m/s². To work out the acceleration of any moving object you need to know two things:

1. The change in velocity.
2. The time taken for this change in velocity.

The acceleration of the object can be calculated using the formula:

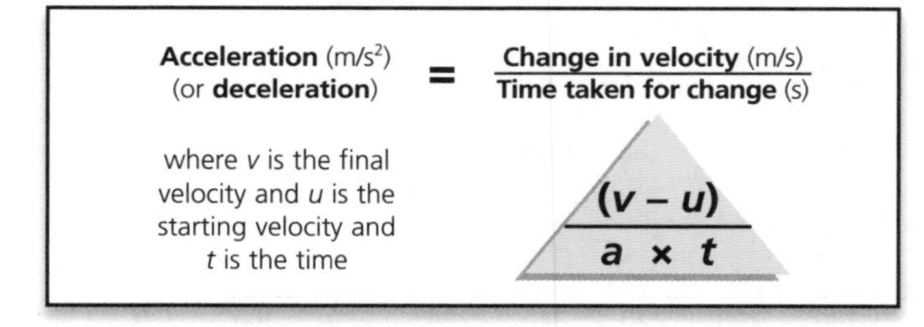

There are two important points to be aware of:

- The cyclist increases his velocity by the *same amount* every second; the *actual* distance travelled each second increases.
- Deceleration is simply a negative acceleration, i.e. it describes an object that is slowing down.

Example

A cyclist accelerates uniformly from rest and reaches a velocity of 10m/s after 5 seconds, then decelerates uniformly and comes to a halt in a further 10 seconds. Calculate **a)** his acceleration, and **b)** his deceleration.

a) Acceleration $= \frac{\text{Change in velocity}}{\text{Time taken}}$

$= \frac{10 - 0}{5} = \mathbf{2m/s^2}$

b) Deceleration $= \frac{\text{Change in velocity}}{\text{Time taken}}$

$= \frac{0 - 10}{10} = \mathbf{-1m/s^2}$

The negative sign shows that it is decelerating. An acceleration of –1m/s² means a deceleration of 1m/s² so the answer should be written as a deceleration of 1m/s².

Velocity–Time Graphs

The slope of a **velocity–time graph** represents the acceleration of the object. The steeper the slope of the graph the greater the acceleration. The area underneath the line in a velocity–time graph represents the total distance travelled.

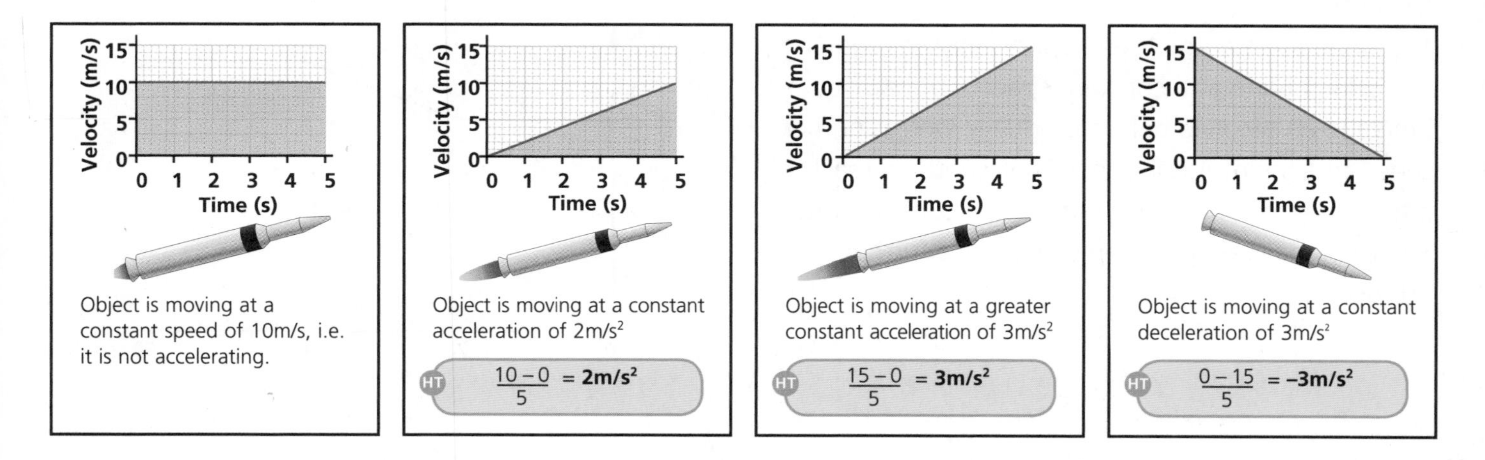

Object is moving at a constant speed of 10m/s, i.e. it is not accelerating.

Object is moving at a constant acceleration of 2m/s²

$\frac{10 - 0}{5} = \mathbf{2m/s^2}$

Object is moving at a greater constant acceleration of 3m/s²

$\frac{15 - 0}{5} = \mathbf{3m/s^2}$

Object is moving at a constant deceleration of 3m/s²

$\frac{0 - 15}{5} = \mathbf{-3m/s^2}$

You need to be able to construct distance–time graphs for a body when the body is stationary or moving in a straight line with constant speed.

Example

An athlete is training for a marathon. She runs at a constant speed for 1 minute and then rests for 20 seconds to allow her pulse rate to slow down. She repeats this four times. For each minute that she is running, she covers 400m. This information can be plotted on a distance–time graph.

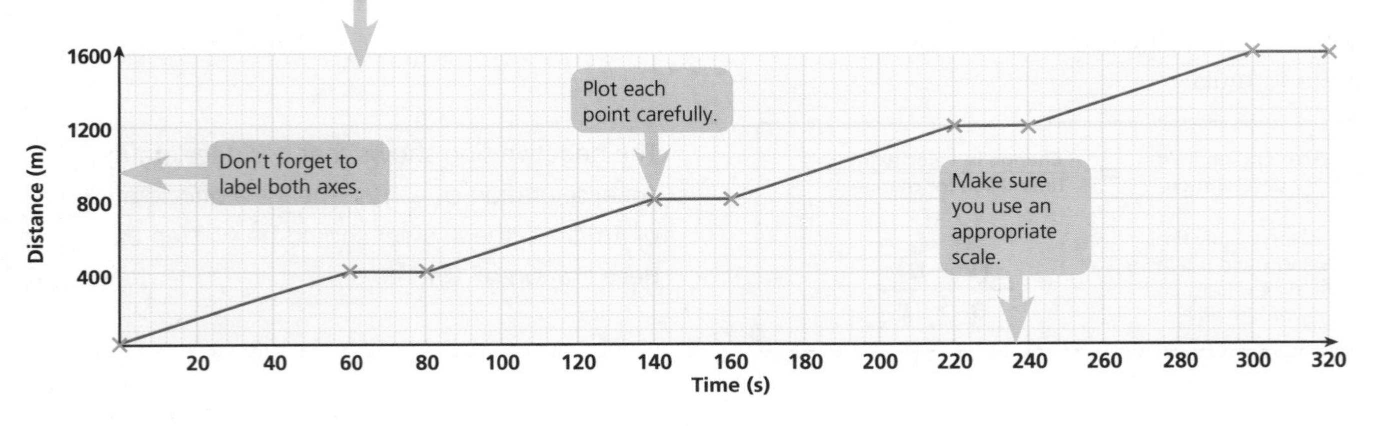

You need to be able to construct velocity–time graphs for a body moving with a constant velocity or a constant acceleration.

Example

The same athlete practises a sprint finish by running in a straight line at a constant speed (i.e. velocity) of 7m/s for 20 seconds before gradually increasing her speed to a sprint. It takes her 20 seconds to reach a top speed of 9m/s accelerating at a constant rate. She then maintains this top speed for a further 20 seconds.

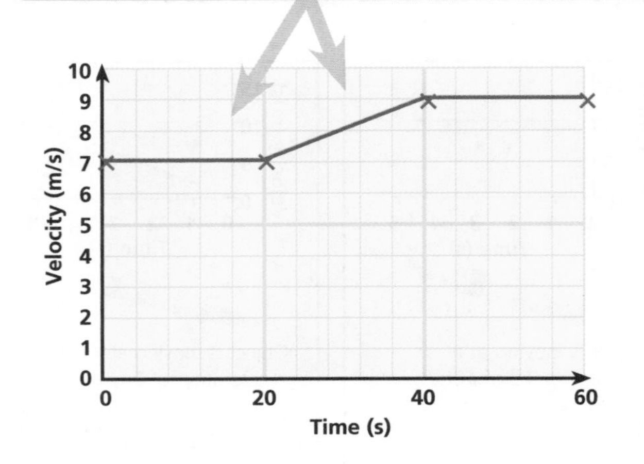

HT **You need to be able to calculate the speed of a body from the gradient of a distance–time graph.**

Example

Here is a distance–time graph. Calculate the speed of the body during the three parts of the journey using the formula:

$$\text{Speed} = \frac{\text{Distance travelled}}{\text{Time taken}}$$

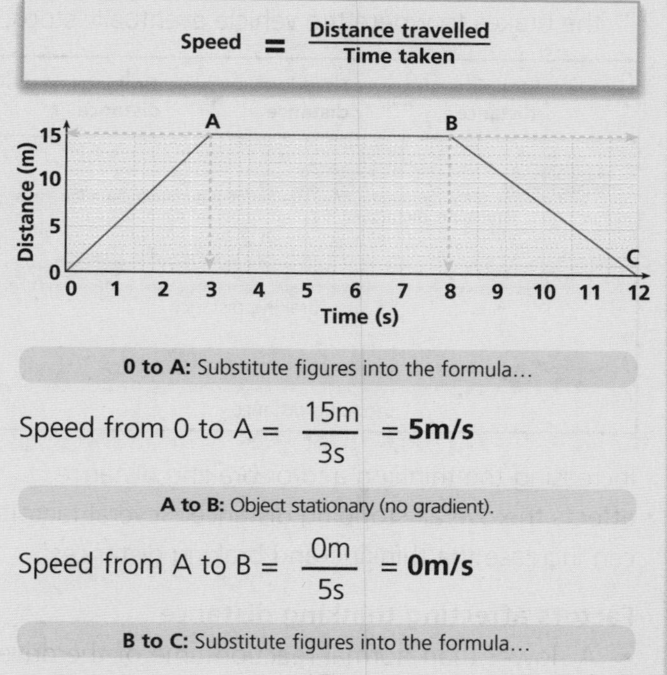

0 to A: Substitute figures into the formula…

Speed from 0 to A = $\frac{15\text{m}}{3\text{s}}$ = **5m/s**

A to B: Object stationary (no gradient).

Speed from A to B = $\frac{0\text{m}}{5\text{s}}$ = **0m/s**

B to C: Substitute figures into the formula…

Speed from B to C = $\frac{15\text{m}}{4\text{s}}$ = **3.75m/s**

So, the object travelled at 5m/s for 3 seconds, remained stationary for 5 seconds, then travelled at 3.75m/s for 4 seconds back to the starting point.

You need to be able to calculate the acceleration of a body from the gradient of a velocity–time graph and the distance travelled from a velocity–time graph.

Example

Here is a velocity–time graph. Calculate the acceleration of the three parts of the journey and the total distance travelled.

$$\text{Acceleration} = \frac{\text{Change in velocity}}{\text{Time taken}}$$

Total distance travelled = Total area under graph

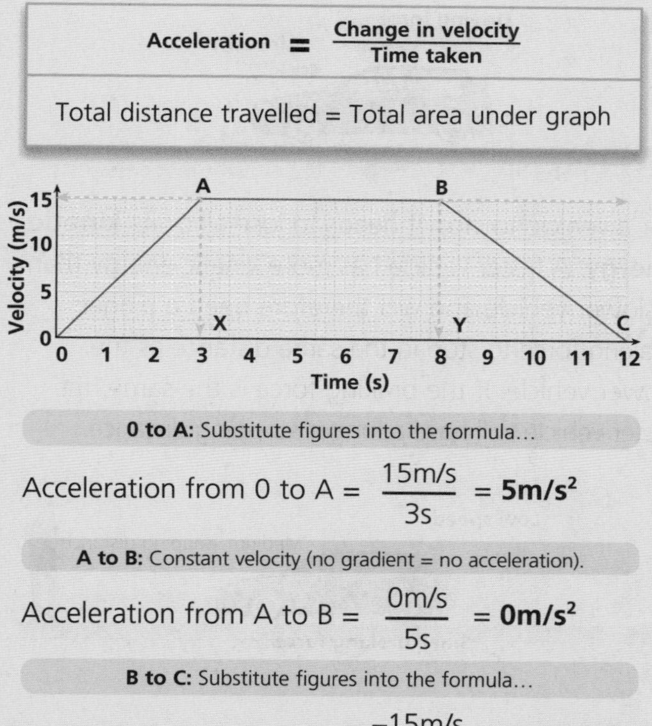

0 to A: Substitute figures into the formula…

Acceleration from 0 to A = $\frac{15\text{m/s}}{3\text{s}}$ = **5m/s²**

A to B: Constant velocity (no gradient = no acceleration).

Acceleration from A to B = $\frac{0\text{m/s}}{5\text{s}}$ = **0m/s²**

B to C: Substitute figures into the formula…

Acceleration from B to C = $\frac{-15\text{m/s}}{4\text{s}}$ = **–3.75m/s²**,

i.e. a deceleration of 3.75 m/s²

So, the object accelerated at 5m/s² for 3 seconds, travelled at a constant speed of 15m/s for 5 seconds before decelerating at a rate of 3.75m/s² for 4 seconds.

The total distance travelled can be calculated by working out the area under the velocity–time graph.

= Area of 0AX + Area of ABYX + Area of BCY

$= (\frac{1}{2} \times 3 \times 15) + (5 \times 15) + (\frac{1}{2} \times 4 \times 15) = \mathbf{127.5m}$

Forces and Braking

You need to know what happens during braking and the factors that affect the stopping distance of a vehicle.

When a car is travelling at constant speed, the driving force from the engine exactly balances the resistive forces. The largest of the resistive forces is air resistance, although there is some friction in the moving parts of the car and rolling resistance in the tyres.

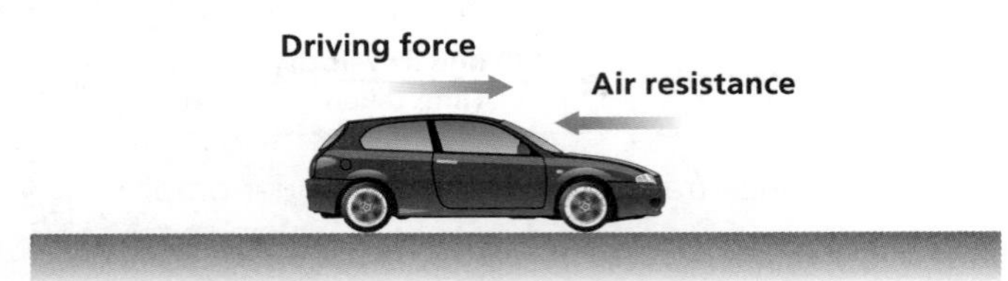

For a vehicle to stop it needs to lose all of its **kinetic energy**. A faster vehicle has more kinetic energy than a slower vehicle and will therefore need a bigger braking force to stop in the same distance as the slower vehicle. If the braking force is the same, the faster vehicle will have a longer stopping distance.

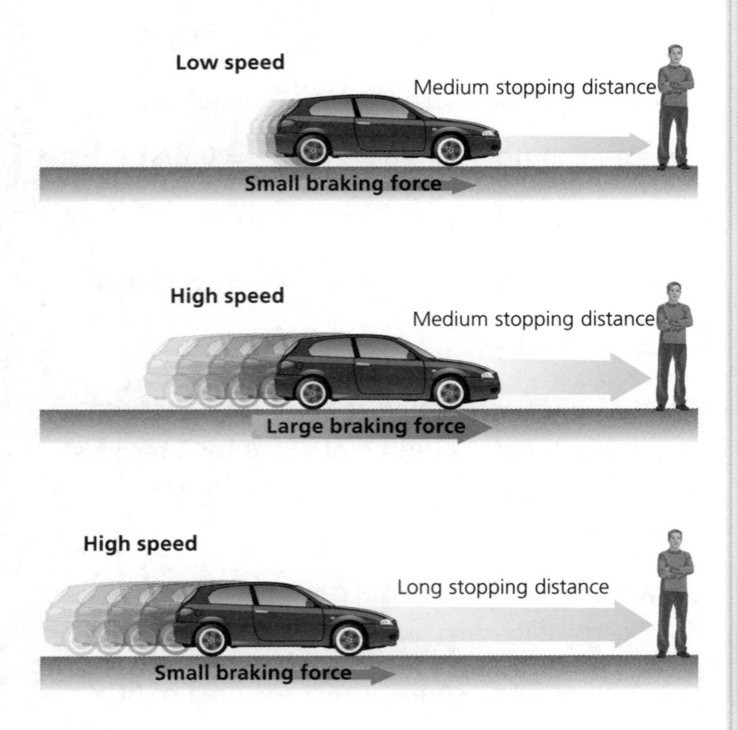

When the brakes are applied, they do work to transfer the kinetic energy of the vehicle into heat energy in the brakes. The brakes, therefore, increase in temperature.

Stopping Distance

The stopping distance of a vehicle depends on:

- **the thinking distance** – the distance travelled by the vehicle from the point when the driver sees that he/she needs to stop to when he/she actually applies the brakes
- **the braking distance** – the distance travelled by the vehicle from the point when the driver applies the brakes to where the vehicle eventually stops.

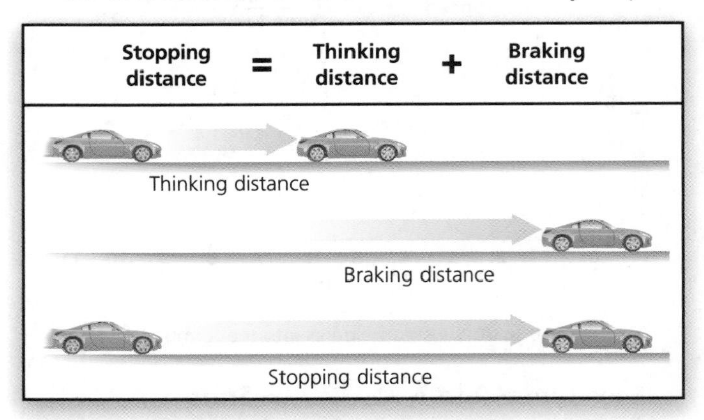

Increasing the thinking and/or braking distance affects the overall stopping distance. Several factors can increase the thinking and braking distances.

Factors affecting thinking distance

- A slower than normal reaction time of the driver, caused by:
  - tiredness or the influence of drugs or alcohol
  - being distracted, e.g. by mobile phone, passengers, etc.

Factors affecting braking distance

- Adverse weather conditions, e.g. wet or icy roads:
  - on a dry road at 50mph the stopping distance is about 50 metres.
  - on a wet or greasy road at 50mph the stopping distance is about 80 metres.
- Poor condition of vehicle, e.g. brakes and tyres.

Factors affecting thinking and braking distance

- The vehicle travelling at greater speeds:
  - at 50mph the stopping distance is about 50 metres (half the length of a football pitch)
  - at 70mph the stopping distance is about 100 metres (the length of a football pitch).

Forces and Terminal Velocity

Falling objects experience two forces:

- the downward force of weight, *W* (↓) which always stays the same
- the upward force of air resistance, *R*, or drag (↑).

When a skydiver jumps out of an aeroplane, the speed of his descent can be considered in two separate parts – **before** and **after** the parachute opens.

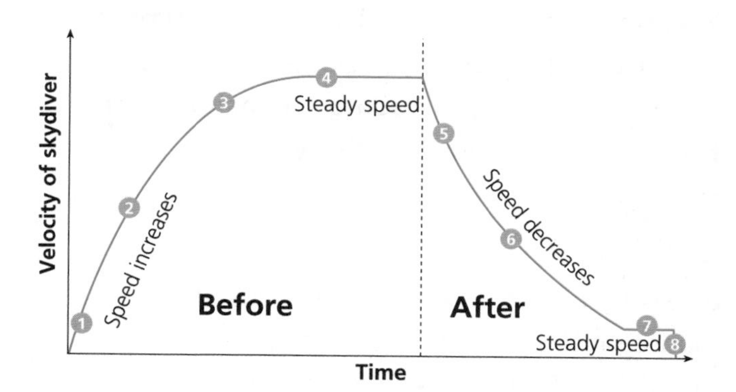

Before the Parachute Opens

When the skydiver jumps, he initially accelerates because of the force due to gravity (1). Gravity is a force of attraction that acts between objects that have mass, e.g. the skydiver and the Earth. The weight of an object is the force exerted on it by gravity. It is measured in newtons (N).

As the skydiver falls he experiences the frictional force of air resistance (*R*) in the opposite direction. But this is not as great as *W* so he continues to accelerate (2).

As his speed increases, so does the air resistance acting on him (3), until eventually *R* is equal to *W* (4). This means that the resultant force acting on him is now zero and his falling speed becomes constant. This speed is called the **terminal velocity**.

After the Parachute Opens

When the parachute is opened, unbalanced forces act again because the upward force of *R* is now greatly increased and is bigger than *W* (5). This decreases his speed and as his speed decreases so does *R* (6).

Eventually *R* decreases until it is equal to *W* (7). The forces are once again balanced and he falls at a steady speed. This new terminal velocity is slower than before because with the parachute opened he is less streamlined.

You need to be able to draw and interpret velocity–time graphs for bodies that reach terminal velocity, including a consideration of the forces acting on the body.

Example

A peregrine falcon is flying above the ground at a velocity of 25m/s. After 10 seconds, it spots a mouse on the ground and goes into a vertical dive making itself into a streamlined shape. It takes 15 seconds for the bird to reach a terminal velocity of 95m/s. It then sustains terminal velocity for 5 seconds. This information can be plotted on a velocity–time graph.

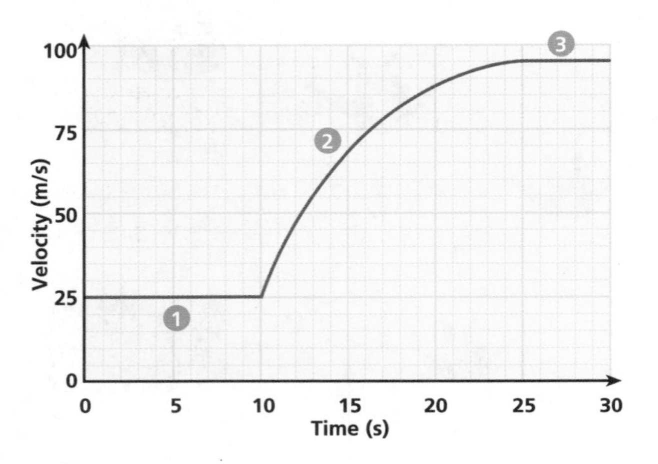

1. The flat line at the start of the graph shows the bird flying at a constant velocity.
2. The curve of the graph begins when the bird enters the dive. The line is steep at first showing a large acceleration. The line gradually becomes less steep because the air resistance acting against the bird decreases the acceleration.
3. The flat line at the top of the graph shows that the bird has reached terminal velocity and is travelling through the air at a constant velocity.

You need to be able to calculate the weight of a body using the formula:

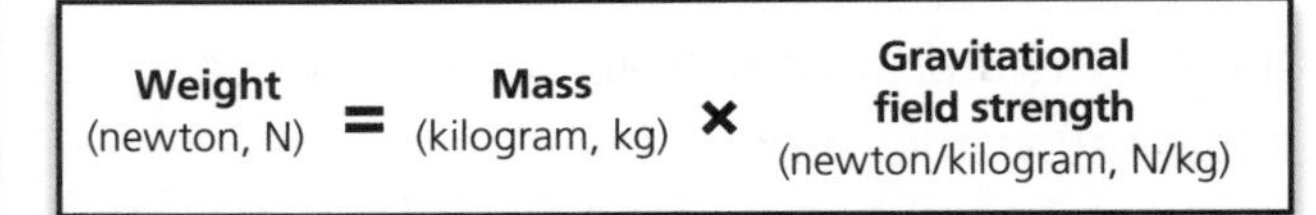

Gravitational field strength is the force that acts on a 1kg mass at a point in a gravitational field. It is measured in newtons per kilogram (N/kg).

Example

The Earth's surface has a gravitational field strength of approximately 10N/kg. The Moon has a gravitational field strength approximately $\frac{1}{6}$ of the Earth's. That means that an object weighs less on the Moon than it does on Earth.

If an astronaut has a mass of 85kg:

- his weight on Earth = 85kg x 10N/kg = **850N**
- his weight on the Moon = 85kg x $\frac{10}{6}$N/kg = **141.7N**

Designing Vehicles

It is not just falling objects that experience drag. All objects travelling in a fluid (liquid or gas) experience a drag force that increases with speed, e.g. cars and submarines.

To increase top speed and be more fuel efficient, vehicles are designed to be streamlined so as to reduce drag. Racing cars are designed to have a powerful engine to give a big driving force while being lightweight and very streamlined / aerodynamic. This combination means that they accelerate quickly and have to be going very fast before the drag force equals the driving force.

Forces and Elasticity

When a **force** is applied to an object it does not always cause it to change speed or direction but may cause it to change its shape instead. Some objects (described as **elastic**) recover their original shape when the force is removed. For these objects, the force makes them stretch or compress and stores elastic potential energy in them.

Examples

A **longbow** – drawing the bow transfers energy from the bowman into the bow, where it is stored as elastic potential energy. When the bow is released, this energy is transferred into the kinetic energy of the arrow.

A **spring jumping toy** – pushing the toy down stores elastic potential energy in the spring. When the toy is released, this energy is transferred into kinetic energy and the toy jumps in the air.

Force and Extension

When a force is used to stretch an elastic object, the **extension** (amount it is stretched by) is directly proportional to the force applied.

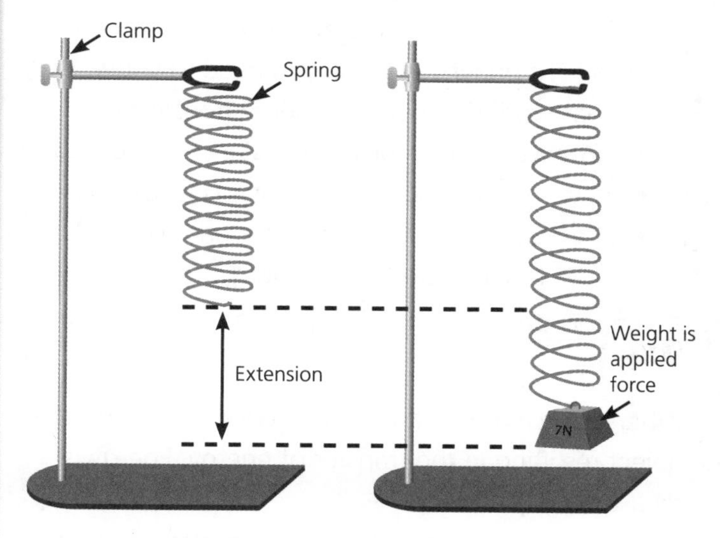

Once the extension exceeds a certain value (called the limit of proportionality) the force and extension stop being directly proportional.

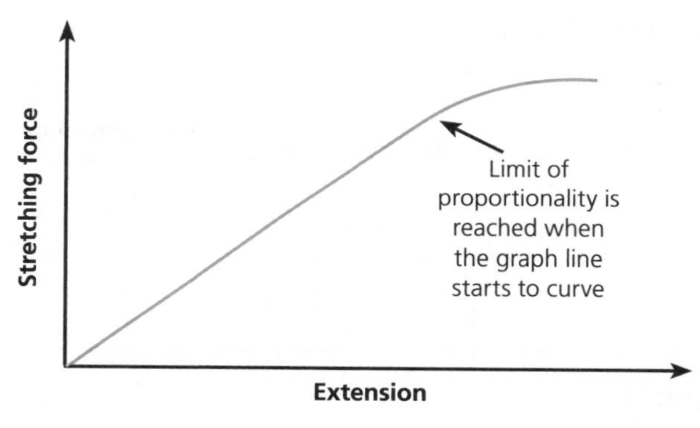

The straight line section of the graph shows where the force and extension are directly proportional. The gradient of this section can be used to find the spring constant (k) with the formula $k = \frac{F}{e}$. This formula is normally written as:

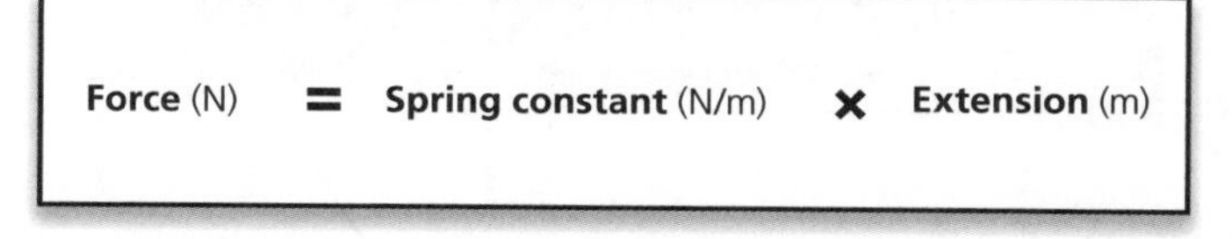

During a straight section on a graph, the material obeys the above formula and is said to obey Hooke's law.

P2.2 The kinetic energy of objects speeding up or slowing down

When an object speeds up or slows down, its kinetic (movement) energy increases or decreases by transferring energy to or from the object. To understand this, you need to know:

- that work done equals energy transferred
- how work done, force and distance are related
- how mass and speed determine kinetic energy
- the result of friction against work done
- how to calculate kinetic energy.

Work

When a force moves an object, **work** is done on the object resulting in the transfer of energy. Energy is measured in **joules** (**J**).

Work done (J) **=** **Energy transferred** (J)

The relationship between work done, force and distance is shown by the formula:

Work done (J) **=** **Force applied** (N) **×** **Distance moved in direction of force** (m)

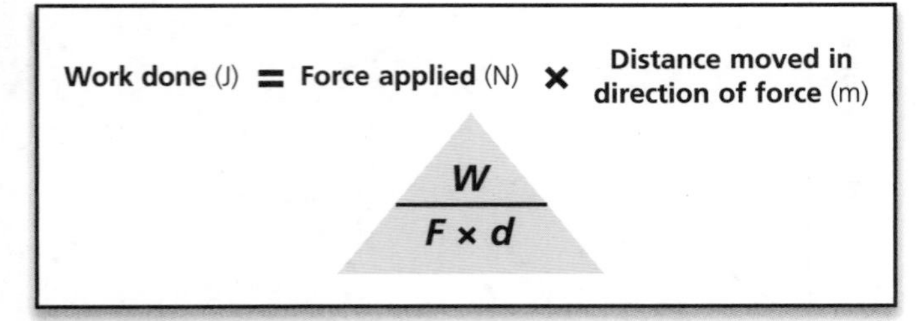

Example

A man pushes a car with a steady force of 250N. The car moves a distance of 20m. How much work does the man do?

Work done = Force applied × Distance moved
= 250N × 20m
= **5000J** (**or 5kJ**)

Work done against frictional forces is mainly transferred into heat energy. For an object that is able to recover its original shape, elastic potential energy is the energy stored in the object when work is done to change its shape.

Kinetic Energy

Kinetic energy is the energy an object has because of its movement. It depends on two things:

- the **mass** of the object (kg)
- the **speed** of the object (m/s).

For example, a moving car has kinetic energy because it has both mass and speed. If the car moves with a greater speed, its mass is unchanged but it has more kinetic energy.

However, a moving truck has a greater mass than the car, so even if it is going slower than the car, it may have more kinetic energy.

Kinetic energy is calculated using this formula:

Kinetic energy (J) **=** $\frac{1}{2}$ **×** **Mass** (kg) **×** **Speed^2**(m/s)^2

KE

$\frac{1}{2} \times m \times v^2$

Example

A car of mass 1000kg is moving at a speed of 10m/s. How much kinetic energy does it have?

Kinetic energy $= \frac{1}{2} \times m \times v^2$

$= \frac{1}{2} \times$ 1000kg × (10m/s)^2

= **50 000J** (**or 50kJ**)

You need to be able to discuss the transfer of kinetic energy to other forms of energy in particular situations.

Example 1: Space shuttles

When a space shuttle returns to Earth it has a lot of kinetic energy, i.e. it has a large mass and is travelling fast. As it enters the Earth's atmosphere, the shuttle encounters frictional forces and kinetic energy is transferred into heat energy, which slows it down.

The shuttle can reach extremely high temperatures because of the heat energy produced, which causes a risk of fire and explosion. Scientists have developed special heat shields to try to protect the body of space shuttles (and the astronauts in them) from this intense heat.

Example 2: Hydroelectricity

Hydroelectric power stations use the kinetic energy in moving water to produce electricity. A dam is built across a river valley and water builds up behind it. The water held behind the dam has gravitational potential energy.

This potential energy is transformed into kinetic energy when the water is released down tubes inside the dam. The moving water is used to drive generators, which transfer the kinetic energy into electrical energy.

Example 3: Bungee rides

One bungee ride consists of a spherical cage that is attached to two supporting arms by elastic cords. Just before the ride starts, the bungee cords are tightened, which provides elastic potential energy.

When the cage (attached to the cords) is released, the elastic potential energy in the cords is transferred into kinetic energy, which propels the cage straight up into the air. The kinetic energy of the cage is reduced as it travels upwards (against the force of gravity, and due to air resistance) and the cords become stretched, so the remaining kinetic energy is transferred into elastic potential energy and gravitational potential energy.

When the cage reaches the top, the elastic and gravitational potential energy is again transferred back into kinetic energy and the cage travels down towards the Earth. The fall is assisted by gravity but the force is again reduced by air resistance. As the cage falls further, kinetic energy is transferred back into elastic potential energy as the cords are stretched. The cage does not return to as low a point as it started because energy has been transferred.

Now, at the bottom of its flight the elastic potential energy is transferred back into kinetic energy. The cage starts to move upwards again and the process is repeated with decreasing amounts of energy each time, until it stops moving.

The Kinetic Energy of Objects

Power, Work and Gravity

When an object is lifted, work is done against the force of gravity. Remember, the weight of an object is the force of gravity on it.

Example

If a weightlifter lifts a 500N weight a distance of 2m, how much work has she done against gravity?

Work done = Force × Distance
= 500 × 2
= 1000J

The speed at which work is done (energy transferred) tells us the power.

$$\textbf{Power} = \frac{\textbf{Energy}}{\textbf{Time}}$$

Example

If the weightlifter takes 2 seconds to lift the 500N weight through a distance of 2 metres, what is the power?

$$\text{Power} = \frac{\text{Energy}}{\text{Time}} = \frac{1000}{2}$$

= **500 watts**

This energy is transferred to the object, which now has an increased amount of gravitational potential energy. The amount of energy gained by the object is the same as the amount of work done in lifting it.

Another way to find the amount of gravitational potential energy gained or lost by an object is to multiply its mass by the change in height and the gravitational field strength.

$$E_p = m \times g \times h$$

where E_p = change in gravitational energy, m = mass, g = gravitational field strength, h = height

The Pendulum

The pendulum is just one example in which gravitational potential energy changes to kinetic energy and back again.

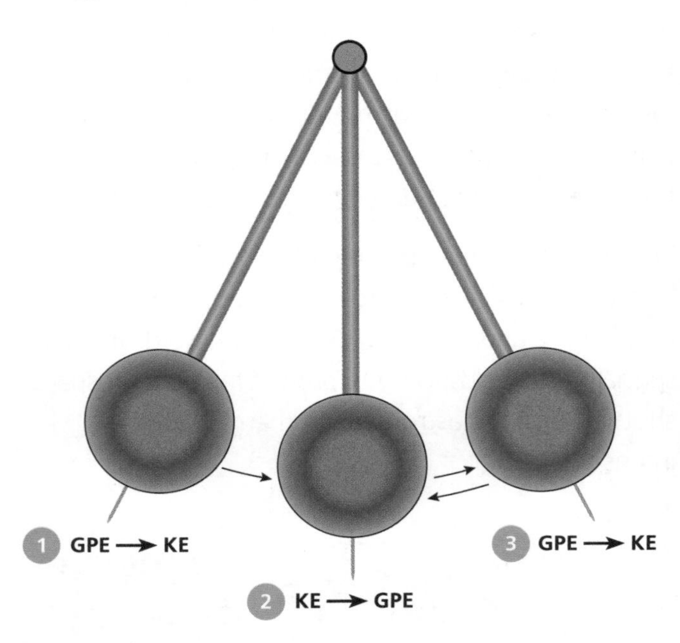

1. Before it is released, the pendulum has gravitational potential energy but no kinetic energy. When it swings downwards, the gravitational potential energy is converted to kinetic energy.
2. All the energy is now kinetic and the pendulum begins to swing back upwards, converting the kinetic energy to gravitational potential energy.
3. All the energy is again gravitational potential energy. The pendulum comes to a stop for a moment as it changes direction. It then swings back down and the energy is again transferred to kinetic energy.

If there were no resistive forces the above process would keep repeating forever as energy is transferred back and forth between the gravitational potential energy and kinetic energy.

In reality, some energy is used to overcome the resistive forces, producing heat. The pendulum comes to a halt after all its energy has been transferred to the surroundings as heat.

What is Momentum?

Momentum is a measure of the state of motion of an object. It depends on two things:

- the **mass** of the object (kg)
- the **velocity** of the object (m/s).

A moving car has momentum because it has both mass and velocity. If the car moves with a greater velocity, it has more momentum providing its mass has not changed. However, a moving truck with a greater mass may have more momentum than the car even if its velocity is less.

The momentum of an object is calculated using the formula:

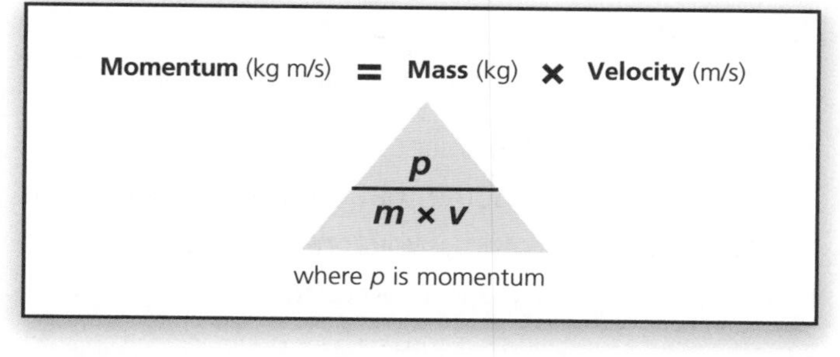

Example 1
A car has a mass of 1200kg. It is travelling at a velocity of 30m/s. Calculate its momentum.

Momentum = Mass × Velocity
= 1200kg × 30m/s
= **36 000kg m/s**

Example 2

A truck has a mass of 4000kg. Calculate its velocity if its momentum is 36 000kg m/s.

Using the formula (rearranged using the formula triangle):

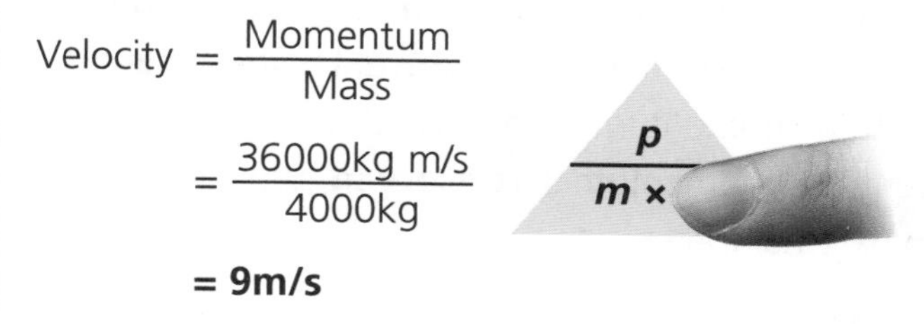

$$\text{Velocity} = \frac{\text{Momentum}}{\text{Mass}} = \frac{36000\text{kg m/s}}{4000\text{kg}}$$

= 9m/s

The truck has a greater mass than the car so it can move at a slower speed and still have the same momentum.

Magnitude and Direction

Velocity and momentum are both quantities that have magnitude (size) and direction.

The direction of movement is especially important when calculating momentum.

For example if car A (which is moving from left to right) has a positive velocity and, consequently, a positive momentum, then car B (moving from right to left) will have a negative velocity and negative momentum because it is moving in the opposite direction to car A.

Example
Car A has a mass of 1000kg and a velocity of 20m/s. Car B has a mass of 1000kg and a velocity of –20m/s. Calculate the momentum of the two cars.

Momentum of car A = 1000kg × 20m/s
= **20 000kg m/s**

Momentum of car B = 1000kg × –20m/s
= **–20 000kg m/s**

Collisions and Explosions

In any collision or explosion, the total momentum in a particular direction after the event is the same as the total momentum in that direction before the event, i.e. momentum is **conserved,** provided that **no** external forces act.

Example 1

Two cars are travelling in the same direction along a road.

Car A collides with the back of car B and they stick together. Calculate their velocity after the collision.

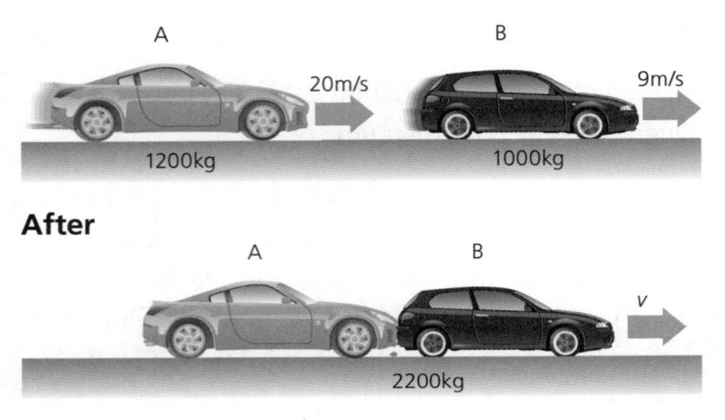

Momentum before collision = Momentum of A + Momentum of B

= (Mass × velocity of A) + (Mass × velocity of B)

= (1200kg × 20m/s) + (1000kg × 9m/s)

= 24 000kg m/s + 9000kg m/s

= **33 000kg m/s**

Momentum after collision = Momentum of A and B stuck together

= (1200 + 1000) × v

= **2200v**

Since momentum is conserved:

Total momentum before = Total momentum after

33 000 = 2200v

Therefore, $v = \frac{33000}{2200}$

= **15m/s**

Example 2

A gun fires a bullet of mass 0.01kg as shown below. The velocity of the bullet is 350m/s. Calculate the recoil velocity of the gun.

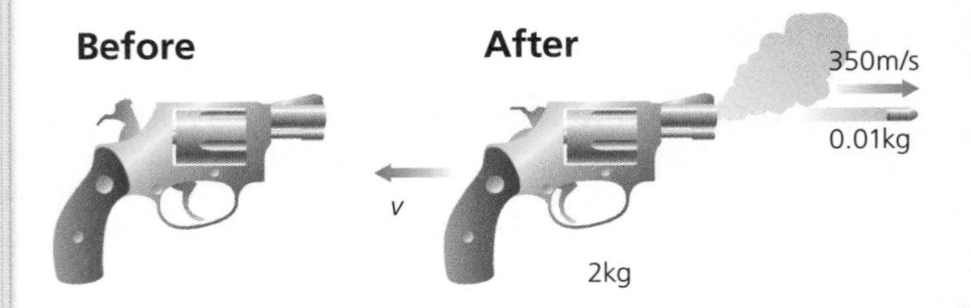

Firing a gun is an example of an explosion where the two objects, i.e. the gun and the bullet, move away from each other, rather than come towards each other, as in a collision.

As we have seen, velocity and momentum are quantities that have magnitude and direction.

Since the gun and the bullet are moving in opposite directions, we will assume that the bullet has positive velocity and momentum, which means that the gun has negative velocity and momentum.

Momentum before explosion = 0

(Neither the gun nor the bullet have momentum because they are not moving.)

Momentum after explosion ...

= Momentum of bullet + Momentum of gun

= (Mass × velocity of bullet) + (Mass × velocity of gun)

= (0.01kg × 350m/s) + (2kg × v)

= 3.5 + 2v

Since momentum is conserved ...

Momentum after explosion = Momentum before explosion

$$3.5 + 2v = 0$$

$$2v = -3.5$$

$$v = \frac{-3.5}{2}$$

$$\mathbf{v = -1.75m/s}$$

Remember, the gun is moving in the opposite direction to the bullet so it has negative velocity.

You need to be able to use ideas of momentum and energy to explain safety and braking features.

Momentum is conserved in a collision. This means that if a vehicle is brought to a sudden halt or makes a sudden change in direction, the people in the vehicle will continue with the same momentum as before unless a force is exerted on them. In other words, any people in the vehicle will continue to travel at the same speed and in the same direction as they were travelling immediately *before the* change.

Obviously, this can have fatal consequences, so cars have inbuilt safety features to try to minimise injury and reduce the number of deaths.

Examples

1. **Seat belts** lock when a car crashes. They exert a force to counteract the momentum of the person wearing them. This prevents the wearer from impacting against the windscreen or smashing against the inside of the car.

   Cars travelling at faster speeds have greater momentum. This means that at high speeds the seatbelt has to exert a larger opposing force, which can result in some bruising and injuries. However, the injuries caused by a seatbelt will be a lot less severe than those it prevents.

2. A **crumple zone** is an area of a car designed to crumple on impact and absorb some of the kinetic energy of the crash. This helps to increase the time over which the car changes momentum, i.e. instead of coming to an immediate halt there will be more time during which the momentum is reduced. This means that the force exerted on the people inside the car will be reduced, which results in fewer injuries.

3. **Power-assisted steering and anti-lock braking systems (ABS)** help the driver to control direction and speed, which can help to reduce the change in momentum.

4. **Air bags** only partially inflate on impact so they are 'squashy'. They distribute the force of impact more evenly over the upper body area and reduce the momentum of the body more gradually.

5. **Side impact bars** are strengthened bars at the side of the car that transmit the forces to the front and the rear of the driver. The kinetic energy of a colliding vehicle is then absorbed in the side impact bars rather than being absorbed by the passengers.

Unlike friction brakes, which transfer kinetic energy to heat and can overheat causing brake failure, **regenerative braking** transfers the kinetic energy of the vehicle to electrical energy. In hybrid or electric vehicles this is used to recharge the batteries and increases the overall efficiency of the vehicle because this energy can be reused rather than lost as heat to the surroundings.

P2.3 Currents in electrical circuits

Static electricity can be explained in terms of electrical charges – when electrical charges move they create an electric current. To understand this, you need to know:

- how materials become electrically charged
- about repulsion and attraction
- how conductors are used
- how electrostatic charges are used.

Static Electricity

Some insulating materials can become electrically charged when they are rubbed against each other. The electrical charge (static) then stays on the material (i.e. it is not discharged).

You can generate static electricity by rubbing a balloon against a jumper. The electrically charged balloon will then attract very small objects.

Static builds up when electrons (which have a negative charge) are 'rubbed off' one material onto another. The material receiving the electrons becomes negatively charged and the one giving up electrons becomes positively charged.

For example, if you rub a Perspex rod with a dry cloth, the Perspex loses electrons to become positively charged. The cloth gains electrons to become negatively charged.

If you rub an ebonite rod with fur, ebonite gains electrons to become negatively charged. Fur loses electrons to become positively charged.

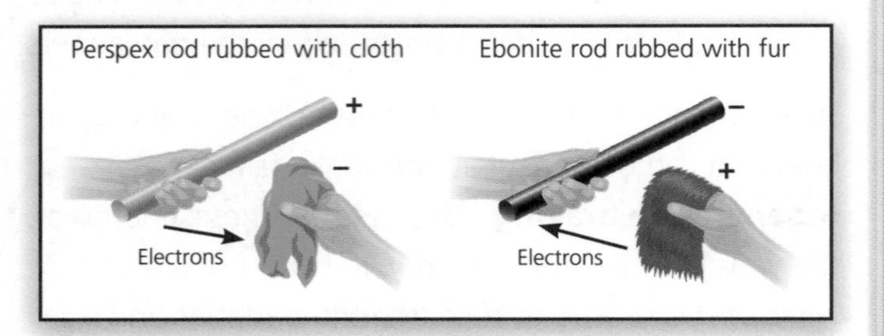

Repulsion and Attraction

When two charged materials are brought together, they exert a force on each other – they are attracted or repelled. Two materials with the same type of charge **repel** each other; two materials with different types of charge **attract** each other.

If you move a Perspex rod near to a suspended Perspex rod, the suspended Perspex rod will be repelled.

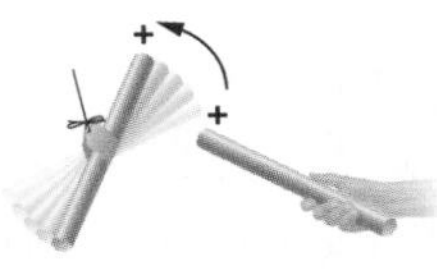

N.B. We would get the same result with two ebonite rods

If you move an ebonite rod near to a suspended Perspex rod, the suspended Perspex rod will be attracted.

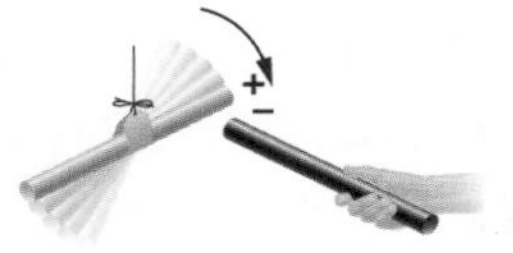

N.B. We would get the same result if the rods were the other way round

Circuit Diagrams

You should know the following standard symbols:

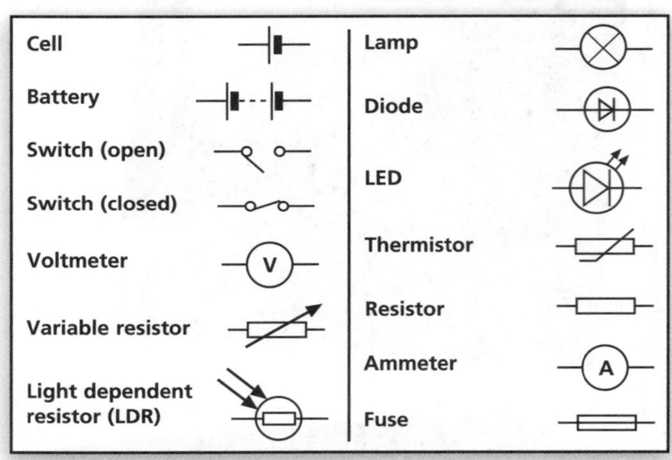

Cell	Lamp
Battery	Diode
Switch (open)	LED
Switch (closed)	Thermistor
Voltmeter	Resistor
Variable resistor	Ammeter
Light dependent resistor (LDR)	Fuse

You need to be able to interpret and draw circuit diagrams using standard symbols.

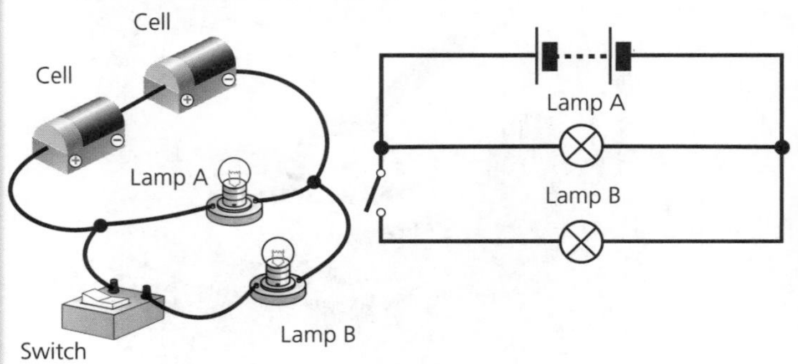

This is a parallel circuit. Lamp A will be on all the time, but lamp B will only come on when the switch is closed. Both lamps will have the same brightness.

Electrical Circuits

Some substances allow electrical charge to flow easily through them. These substances are called **conductors**. Electric current is the flow of charge. The quicker the charge moves the greater the current.

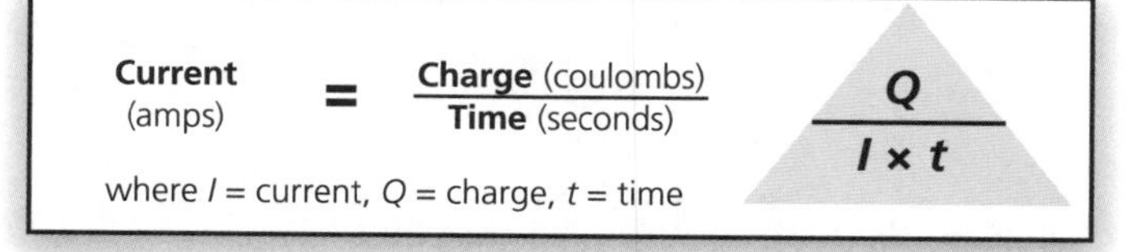

$$\text{Current (amps)} = \frac{\text{Charge (coulombs)}}{\text{Time (seconds)}}$$

where I = current, Q = charge, t = time

An electric current will flow through a component if there is a potential difference (voltage) across the ends of the component.

Circuit 1

Cell provides p.d. across the lamp. A current flows and the lamp lights up.

The amount of current flowing depends on the resistance of the bulb and the potential difference.

Potential Difference

The **potential difference** tells us how much work is done (energy transferred) per coulomb of charge that flows through a component.

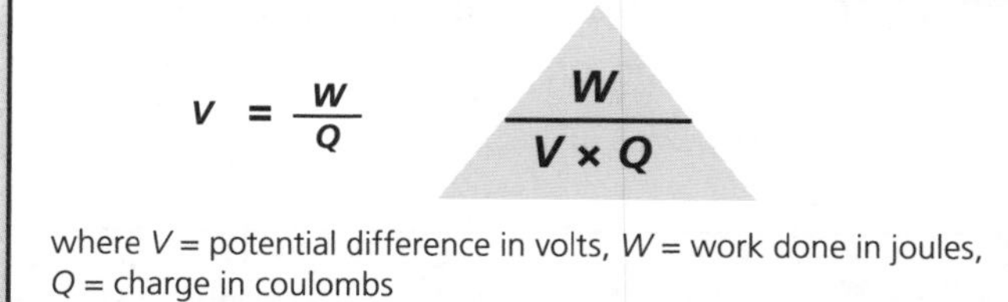

$$V = \frac{W}{Q}$$

where V = potential difference in volts, W = work done in joules, Q = charge in coulombs

In practice this means that a bigger potential difference produces a bigger current and provides more energy. Therefore, the bulbs will be brighter.

Circuit 2

Two cells together provide a bigger p.d. across the lamp. A bigger current now flows and the lamp lights up more brightly (compared to circuit 1).

The Resistance of the Component

Components resist the flow of current through them, i.e. they have **resistance** (measured in ohms).

The greater the resistance of a component or components, the smaller the current that flows for a particular voltage, or the greater the voltage needed to maintain a particular current.

Circuit 3

Two lamps together have a greater resistance. A smaller current now flows and the lamps light up less brightly (compared to circuit 1).

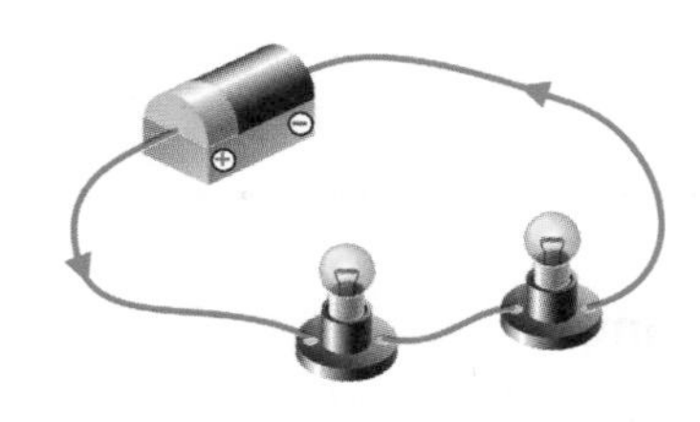

HT A good conductor has a large number of free electrons so has a low resistance. A component with a high resistance will resist the flow of these electrons.

As the temperature of most components increases, the ions in the metal vibrate faster and get in the way of the electrons. This increases the resistance.

Potential Difference and Current

The potential difference across a component in a circuit is measured in **volts (*V*)** using a **voltmeter** connected in **parallel** across the component.

The current flowing through a component in a circuit is measured in **amperes (*A*)**, using an **ammeter** connected in **series**.

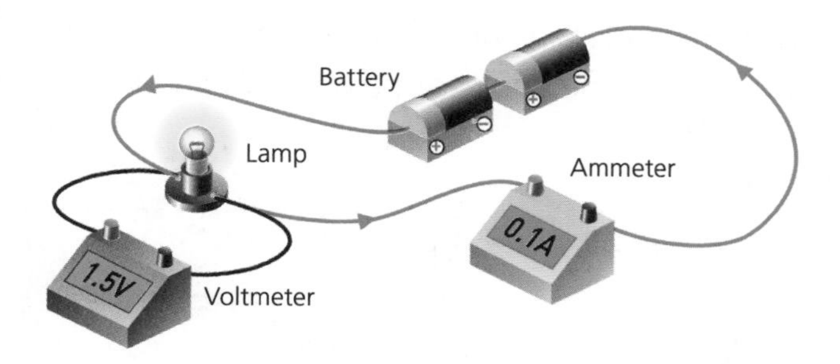

Currents in Electrical Circuits

Resistance

Resistance is a measure of how hard it is to get a current through a component at a particular potential difference (voltage). Potential difference, current and resistance are related by the formula:

Potential difference (volt, V)	=	Current (ampere, A)	×	Resistance (ohm, Ω)

$$\frac{V}{I \times R}$$

where I is current

Example

Calculate the reading on the voltmeter in this circuit if the bulb has a resistance of 15 ohms.

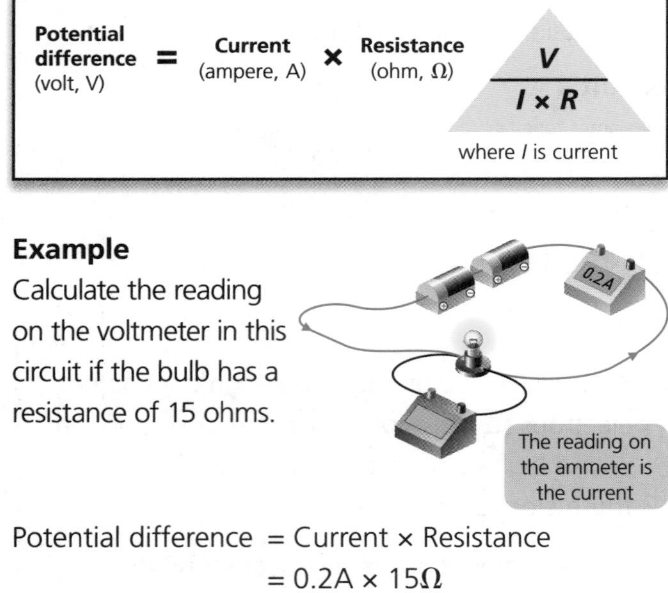

Potential difference = Current × Resistance
= 0.2A × 15Ω
= 3V

Resistance of Components

The resistance of components can be investigated using a voltmeter, ammeter and a variable power supply. You can draw **current–potential difference graphs** that show how the current through the component varies with the voltage across it.

Resistors

As long as the temperature of the resistor stays constant, the current through the resistor is directly proportional to the potential difference across the resistor, regardless of which direction the current is flowing, i.e. if one doubles, the other also doubles.

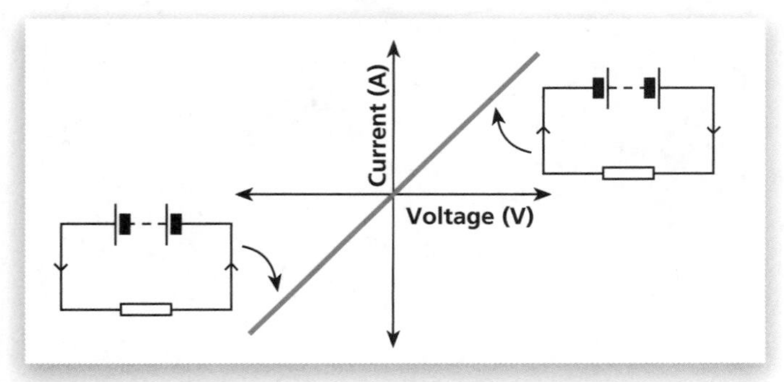

Filament Lamps

As the temperature of the filament increases and the bulb gets brighter then the resistance of the lamp increases, regardless of which direction the current is flowing.

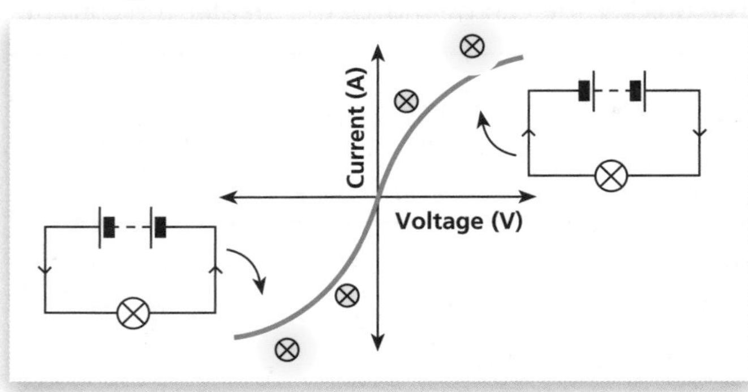

Diodes

A diode allows a current to flow through it in one direction only. It has a very high resistance in the reverse direction so no current flows.

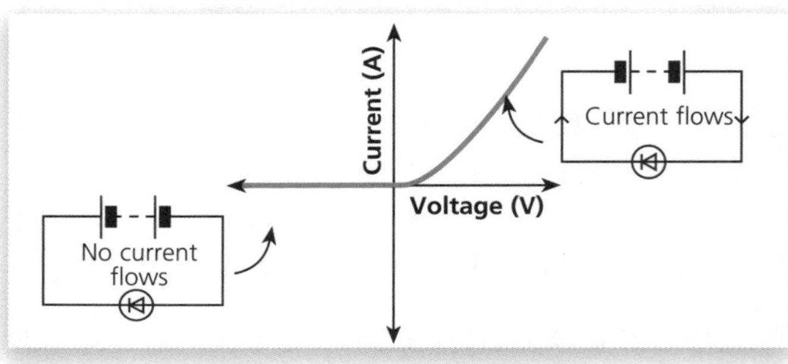

Light Dependent Resistor (LDR)

The resistance of an LDR depends on the amount of light falling on it. Its resistance decreases as the amount of light falling on it increases; this allows more current to flow.

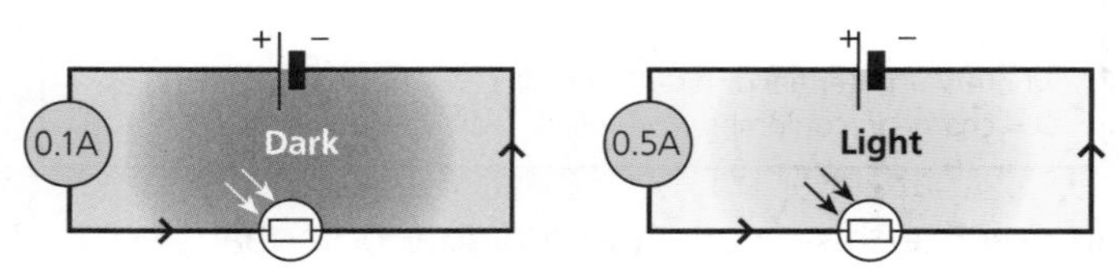

Thermistor

The resistance of a thermistor depends on its temperature. Its resistance decreases as the temperature of the thermistor increases; this allows more current to flow.

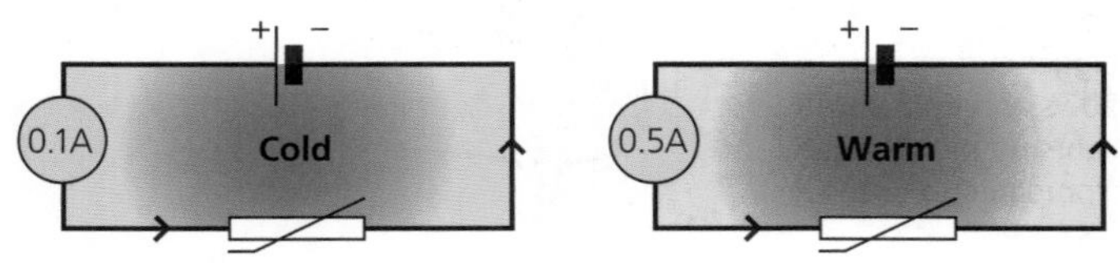

Series and Parallel Circuits

Components Connected in Series

In a series circuit, all the components are connected one after the other in one loop, going from one terminal of the battery to the other.

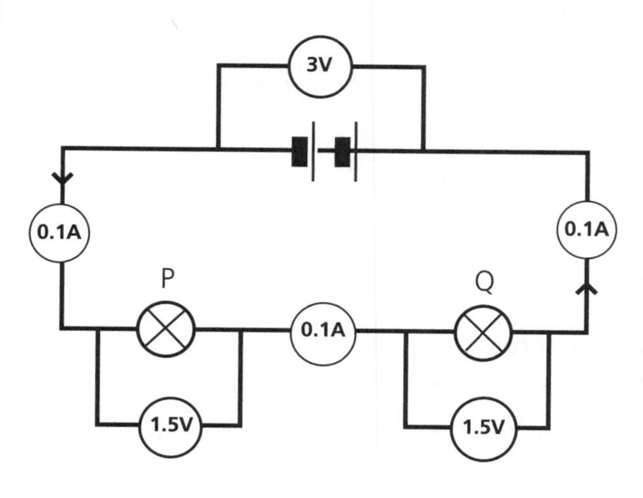

- The same current flows through each component. In the circuit above, each ammeter reading is 0.1A.
- The potential difference (voltage) supplied by the battery is divided up between the components in the circuit. In the circuit above both bulbs have the same resistance so the potential difference is divided equally. If one bulb had twice the resistance of the other, then the potential difference would be divided differently, 2V and 1V.
- The total resistance is the sum of the individual resistances of the components, i.e. $R = R_P + R_Q$. In the circuit above, if both P and Q each have a resistance of 15 ohms, the total resistance would be 15 ohms + 15 ohms = 30 ohms.

Components Connected in Parallel

Components connected in parallel are connected separately in their own loop going from one terminal of the battery to the other.

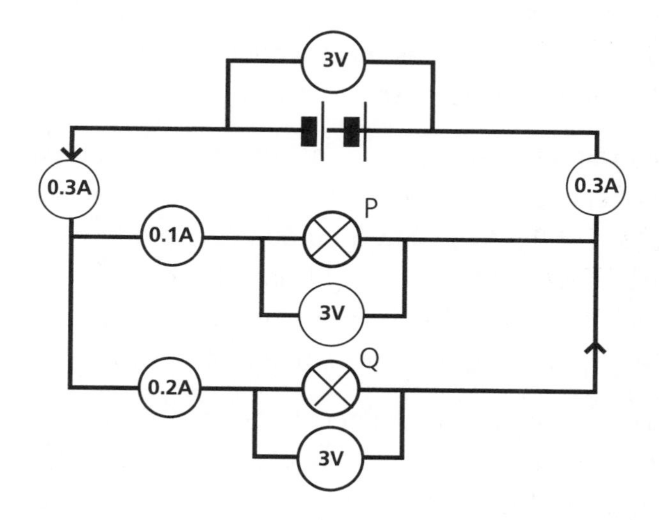

- The total current in the main circuit is equal to the sum of the currents through the separate components. In the circuit above, 0.3A = 0.1A + 0.2A = 0.3A.
- The potential difference across each component is the same (and is equal to the p.d. of the battery). In the circuit above, each bulb has a p.d. of 3V across it.
- The amount of current that passes through a component depends on the resistance of the component. The greater the resistance, the smaller the current. In the circuit above, bulb P must have twice the resistance of bulb Q because only 0.1A passes through bulb P while 0.2A passes through bulb Q.

Connecting Cells in Series

The total potential difference provided by cells connected in series is the sum of the p.d. of each cell separately, providing that they have been connected in the same direction.

Each of the following cells has a p.d. of 1.5V.

Total p.d. of battery
= 2 × 1.5V
= **3V**

Total p.d. of battery
= 3 × 1.5V
= **4.5V**

You need to be able to evaluate the use of different forms of lighting, in terms of cost and energy efficiency.

Not all devices with the same power rating will produce the same effect because it depends how efficient a device is at transferring energy usefully. As charge flows through a resistor, the resistor gets hot, wasting energy. The less heat produced, the more efficient the device is likely to be.

For example, filament lamps get hot, which wastes a lot of energy compared to compact fluorescent bulbs (CFLs). For the same power output, CFLs produce around five times the light of filament lamps. In reality this means that CFLs use only a fifth of the energy producing the same amount of light.

LEDs are also used for lighting because of their low temperature and very low running costs. An LED is a diode, so when connecting one it is important to put it in the correct way around in the circuit because it only allows current to flow in the forward direction. Unfortunately LEDs do not give out light evenly in all directions. This makes them cost effective as spot lights but not as useful as room lights where a large number would be needed. So the cost would outweigh the efficiency saving.

When comparing possible solutions you need to look at purchase price and efficiency saving to decide which is the most cost effective.

You need to be able to apply the principles of basic electrical circuits to practical solutions.

Temperature Warning Lights
A simple use of thermistors is on temperature indicator lights. As the temperature rises, the thermistor resistance falls and more current flows. The light then switches on.

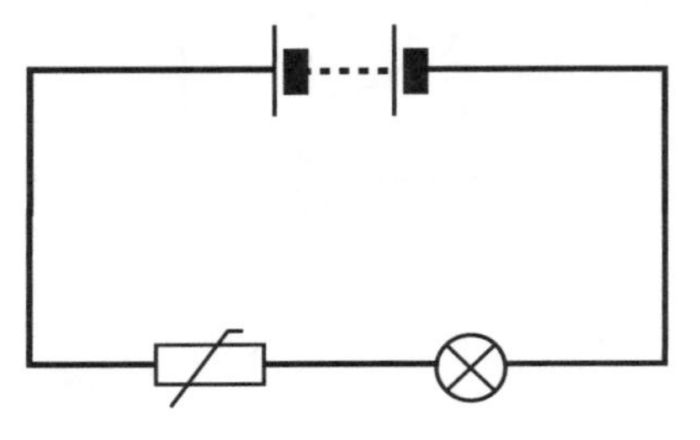

Thermostats
A thermostat uses a thermistor to turn on a heater when the temperature drops below a certain level.

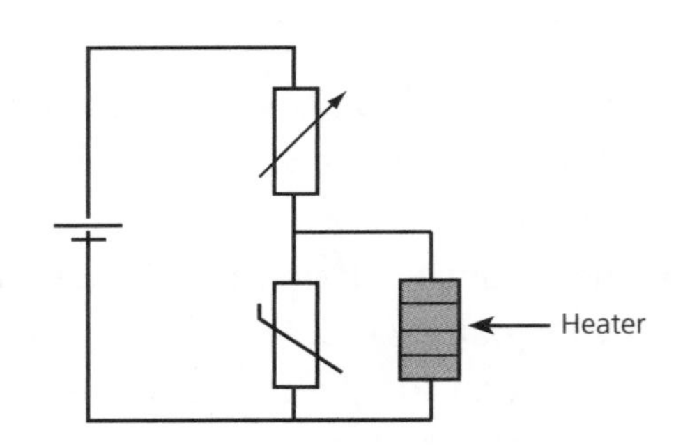

At a low temperature the thermistor has a high resistance, and this forces the current through the heater, switching the heater on. As the temperature increases, the resistance of the thermistor falls so the current goes through the thermistor instead of the heater, switching the heater off. Adjusting the variable resistor changes how much current flows through the heater and can be used to set the temperature.

Automatic Lights
Automatic lights using light dependent resistors work in exactly the same way as above but they use an LDR and a light instead of a thermistor and a heater.

When dark, the LDR has a high resistance so the current flows through the light bulb, switching it on. In bright light, the LDR has low resistance so the current flows through the LDR and not the bulb.

P2.4 Using mains electricity safely and the power of electrical appliances

Although useful, mains electricity can be very dangerous so it is important to know how to use it safely. To understand this, you need to know:

- what a direct current is
- what an alternating current is
- how electrical appliances are connected to the mains
- the structure and wiring of a three-pin plug
- the use of a circuit breaker, fuse and the earth wire.

Currents

A **direct current** (**d.c.**) always flows in the same direction. Cells and batteries supply direct current.

An **alternating current** (**a.c.**) changes the direction of flow back and forth continuously. The number of complete cycles of reversal per second is called the **frequency**, and for UK mains electricity this is 50 cycles per second (Hertz).

In the UK, the mains supply has a voltage of about 230 volts, which can kill if it is not used safely.

The Three-pin Plug

Most electrical appliances are connected to the mains electricity supply using a cable and a three-pin plug that is inserted into a socket on a ring main circuit.

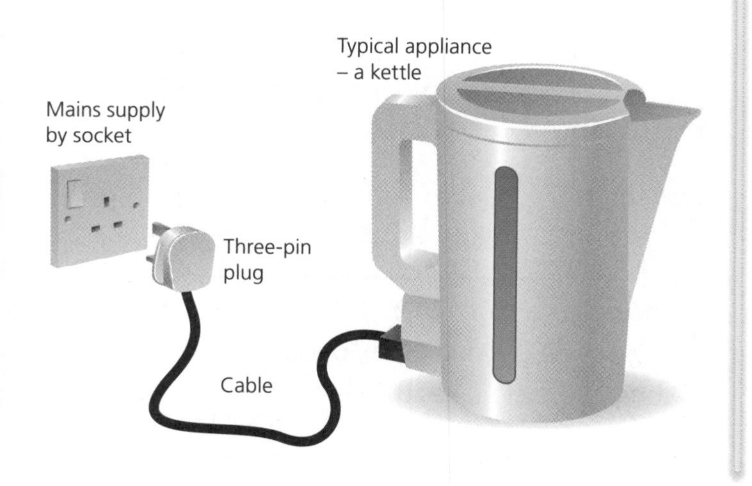

Earth wire (green and yellow) – all appliances with outer metal cases are earthed

Live wire (brown) – carries current to the appliance

Fuse – always part of the live circuit. Should be of the proper current rating

5A

Neutral wire (blue) – carries current away from appliance

Casing – plastic or rubber because both are good insulators

Cable grip – secures the cable in the plug

Cable

- The inner cores of the wires are made of copper because it is a good conductor.
- The outer layers are made of flexible plastic because it is a good insulator.
- The pins of a plug are made from brass because it is a good conductor.
- Some appliances are double insulated, which means that no single fault can lead to an exposed live connection. These appliances do not have an earth connection.

You need to be able to compare and calculate potential differences (voltage) of d.c. supplies and the peak potential differences of a.c. supplies from diagrams of oscilloscope traces.

Frequency and voltage can be compared using an oscilloscope. The diagrams below have 1V per division.

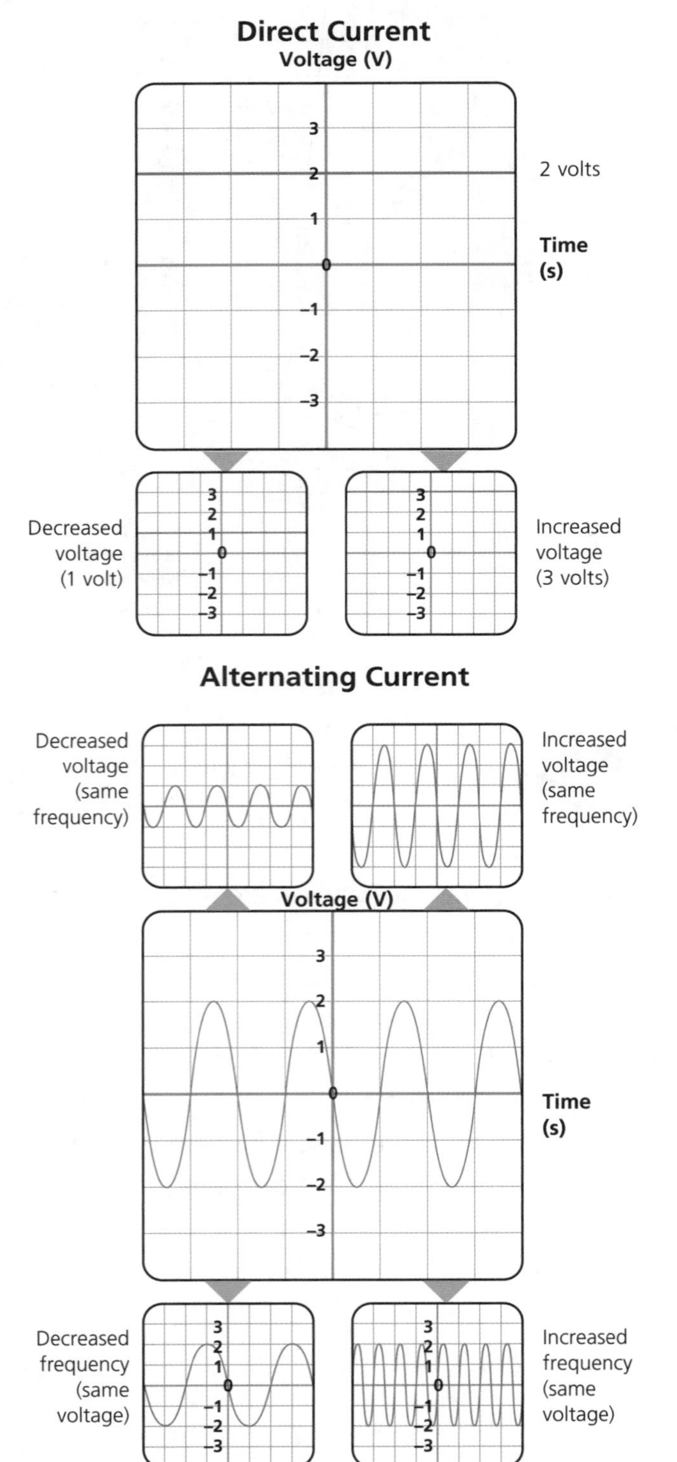

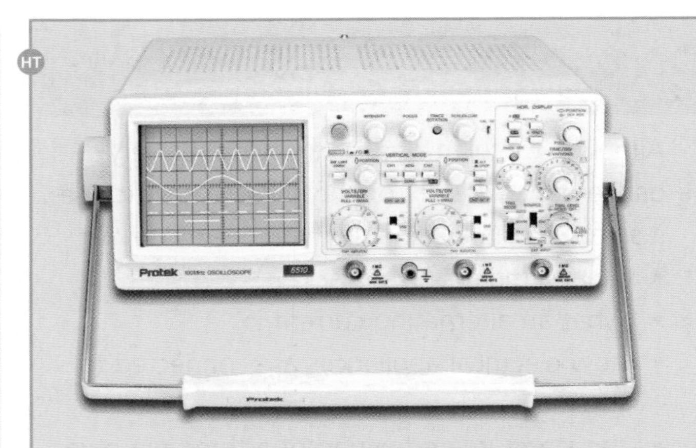

You need to be able to determine the period, and hence the frequency, of a supply from diagrams of oscilloscope traces.

Example

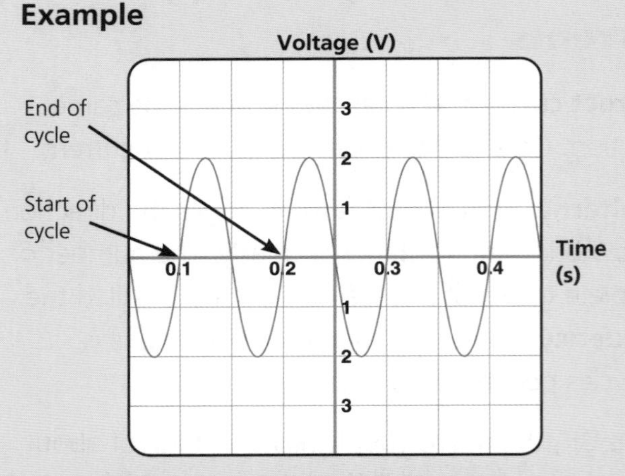

The peak voltage is 2 volts and the frequency is 10 cycles per second (10Hz), because one complete cycle takes 0.1 seconds.

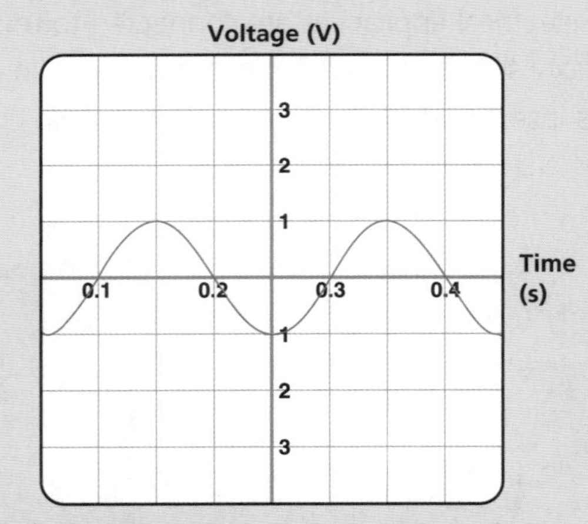

The peak voltage is 1 volt and the frequency is 5 cycles per second (5Hz), because one complete cycle takes 0.2 seconds.

Circuit Breakers and Fuses

A **residual current circuit breaker** (**RCCB**) is a device that automatically breaks an electric circuit if the circuit becomes overloaded. It does so by detecting a difference between the live and the neutral wires that should not be there.

When an RCCB detects a fault, it trips a switch to break the circuit much quicker than a fuse would. This makes an RCCB better at protecting users. RCCBs are often used with lawnmowers and hedge trimmers where there is a danger of the cable being cut. They are also used with computers to protect the computer from power surges, etc.

A Circuit Breaker

current becomes too high

RCCB trips switch

circuit is broken

cable or appliance is protected

A **fuse** is a short, thin piece of wire with a low melting point. When the current passing through it exceeds the current rating of the fuse, the fuse wire gets hot and melts or breaks, breaking the circuit. This prevents damage to the cable or the appliance that would be caused by overheating.

The current rating of the fuse must be just above the normal working current of the appliance for the safety system to work properly. A thicker fuse wire has a higher fuse value.

A Fuse

current larger than current rating of fuse

fuse wire melts

circuit is broken

cable or appliance is protected

Earthing

All electrical appliances with outer metal cases must be earthed. The outer case of an appliance is connected to the earth pin in the plug by the earth wire.

If a fault in the appliance connects the live wire to the case, the case will become **live.** The current will then 'flow to earth' through the earth wire because this offers least resistance. This overload of current will cause the fuse to melt, or the circuit breaker to trip. Therefore, the earth wire and fuse work together to protect the appliance and user.

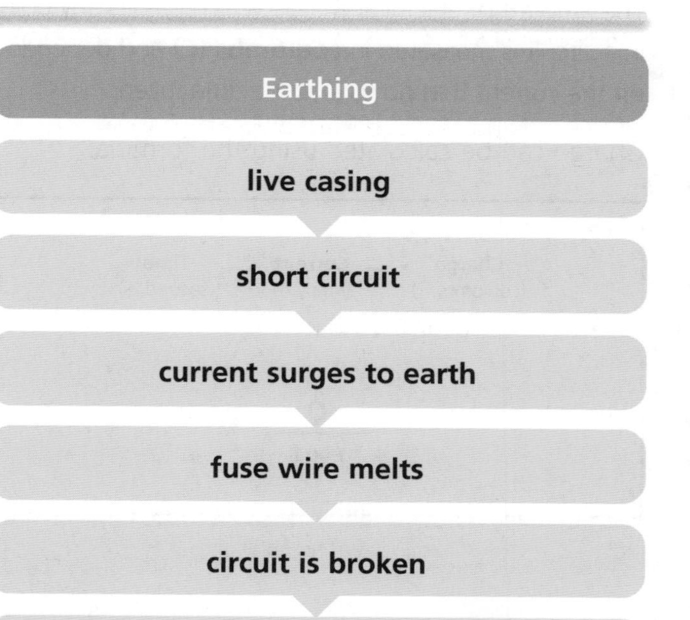

Power

An electric **current** is the rate of flow of charge – it **transfers** electrical energy from a battery or power supply to components in a circuit. The rate of flow is measured in **amperes (A)**.

The components transform some of this electrical energy into other forms of energy, e.g. a resistor **transfers** electrical energy into heat energy. The rate at which energy is transferred in a device is called the **power**.

This can be calculated using the formula:

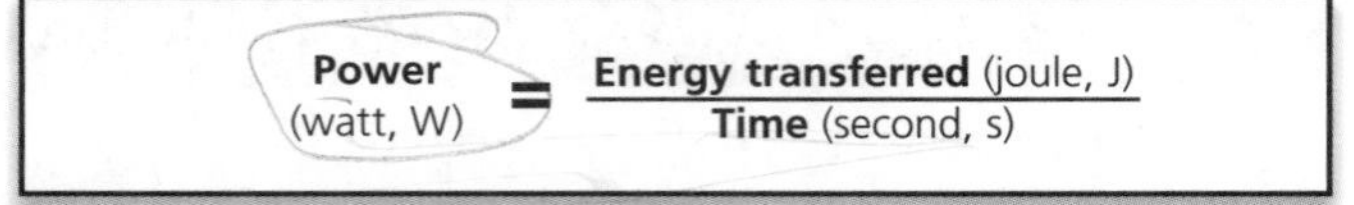

Power can also be calculated using the formula:

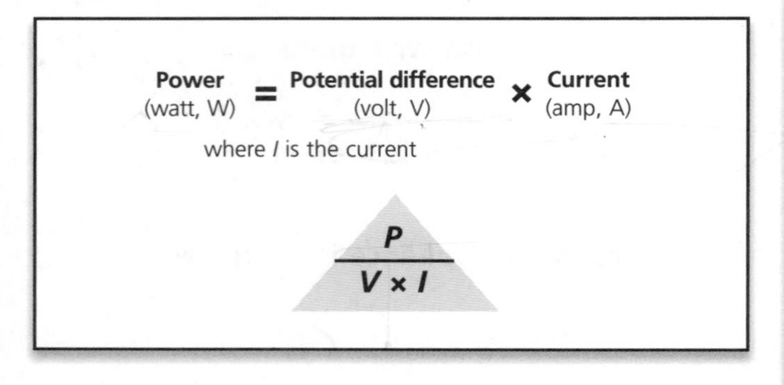

Charge

The amount of electrical charge which passes any point in a circuit is measured in **coulombs (C)** and depends on the current that flows and the time taken.

Charge can be calculated using the formula:

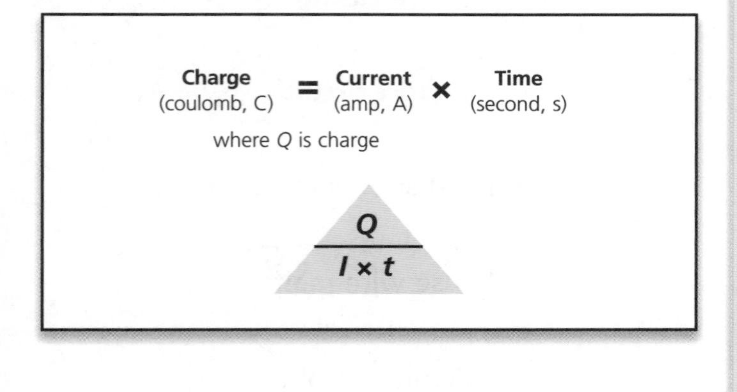

Example

If the circuit below is switched on for 40 seconds and the current is 0.5 amps, how much charge flows?

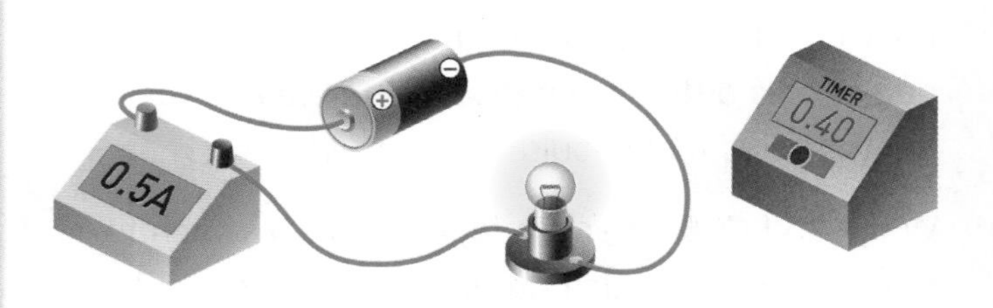

Charge = Current x Time
= 0.5A x 40s = **20 coulombs**

HT Transferring Energy

As charge passes through a device, energy is transferred. The amount of energy transferred by every coulomb of charge depends on the size of the potential difference.

The greater the potential difference, the more energy transferred by every coulomb of charge.

Energy transferred, potential difference and charge are related by the formula:

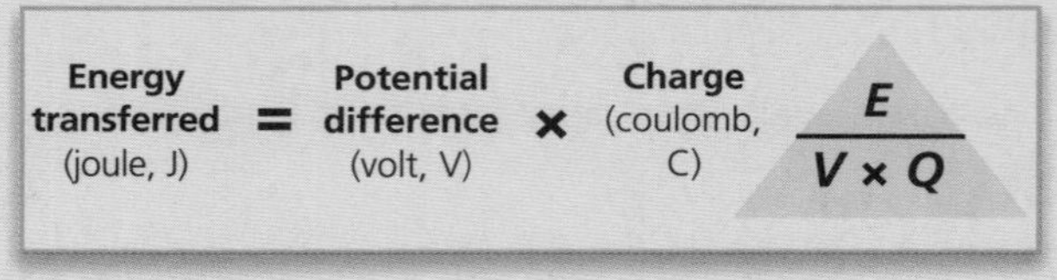

Example

If a circuit has a potential difference of 1.5V and passes a charge of 20C, how much energy is transferred?

Energy transferred = p.d. × Charge
= 1.5V × 20C
= **30 joules**

Remember, the charge gained this energy from the supply voltage, i.e. the battery, which it transferred to the bulb in the 40 seconds the circuit was switched on.

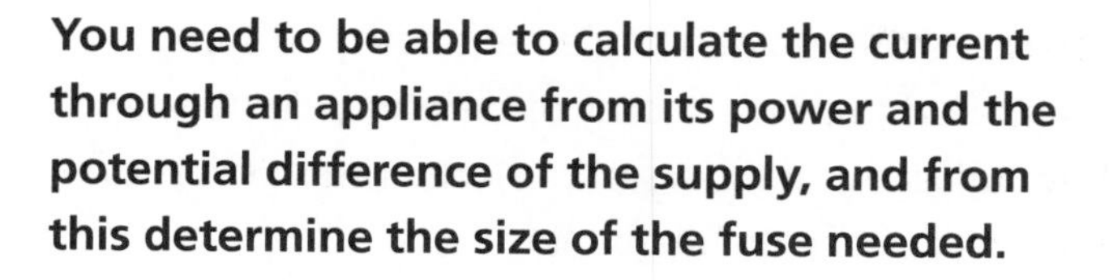

You need to be able to calculate the current through an appliance from its power and the potential difference of the supply, and from this determine the size of the fuse needed.

When a new appliance is developed, the manufacturer needs to work out what size fuse it requires.

If the power and operating voltage (potential difference) are known, the current can be calculated by rearranging the power formula:

$$\text{Current} = \frac{\text{Power}}{\text{Potential difference}}$$

Once the current has been calculated, the size of the fuse can then be determined. The fuse needs to be as close to the current as possible, but higher than it.

Fuses come in the following standard sizes: 1 amp, 3 amp, 5 amp, 13 amp, 20 amp, 25 amp and 30 amp.

Example

Device	Power	Voltage	Current	Ideal Fuse
Television	60W	230V	$\frac{60}{230}$ **= 0.26A**	1 amp
Oven	5500W	230V	$\frac{5500}{230}$ **= 23.9A**	25 amp
Computer	43W	16V	$\frac{43}{16}$ **= 2.7A**	3 amp
Drill	800W	230V	$\frac{800}{230}$ **= 3.5A**	5 amp
Microwave	1150W	230V	$\frac{1150}{230}$ **= 5A**	13 amp

Safety Precautions

Mains electricity can be deadly, so we need to ensure safe practice when using mains appliances.

For example:

- electrical sockets are not allowed in bathrooms and should not be situated close to water supplies
- connecting multi-socket extension leads together can overload the electrical system, which (if the fuse does not work properly) could result in a fire.

When choosing cables for different appliances the same considerations have to be made as with fuses. For example, using a 5 amp cable for a 13 amp appliance will melt the wires in it and could set fire to the insulation around it. The cable should be able to carry the maximum current without heating up.

In some cases cables should have a heat resistant or toughened covering if they are going to be exposed to potential damage (e.g. irons and vacuum cleaners).

P2.5 What happens when radioactive substances decay, and the uses and dangers of their emissions

We need to understand the structure of atoms in order to understand what happens to radioactive substances when they decay. To understand this, you need to know:

- the relative masses and relative charges of protons, neutrons and electrons
- what ions are
- what isotopes are
- what the mass number represents
- what effect alpha and beta decay have on radioactive nuclei
- the origins of background radiation.

Atoms

An atom is made up of three parts:

- **protons**
- **neutrons**
- **electrons**.

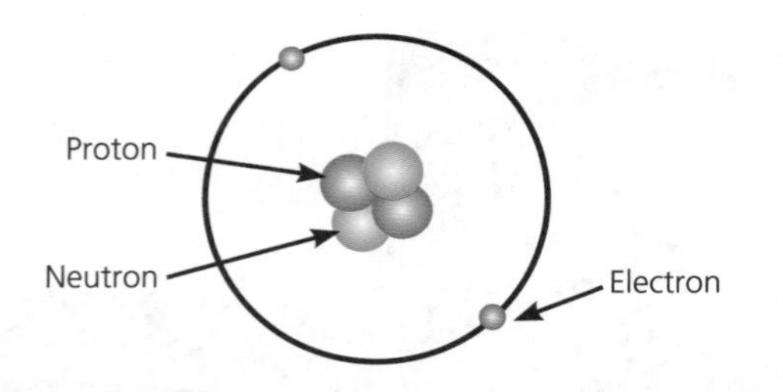

Atomic Particle	Relative Mass	Relative Charge
Proton	1	+1
Neutron	1	0
Electron	0 (nearly)	–1

Normally an atom has the same number of protons as electrons, so atoms have no electrical charge. Atoms can gain or lose electrons. If an atom gains electrons it becomes a negative ion; if it loses electrons it becomes a positive ion.

All atoms of a particular element have the same number of protons. Atoms of different elements have different numbers of protons. The number of protons defines the element.

The total number of protons and neutrons in an atom is called its **mass number**. The number of protons in an atom is called its **atomic number**.

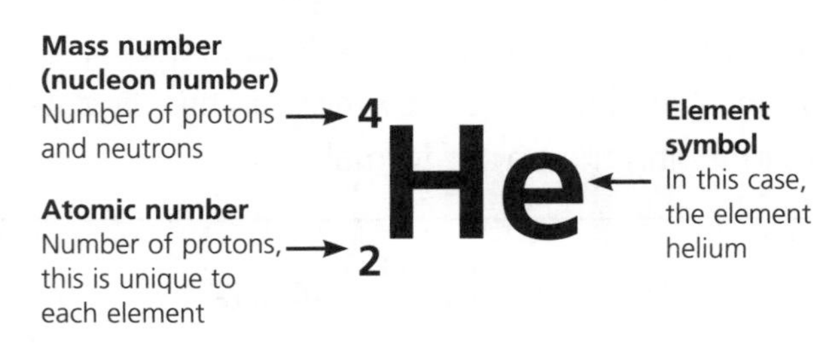

However, atoms of the same element can have different numbers of neutrons. These are called **isotopes**.

Oxygen has several isotopes, three of which are shown here:

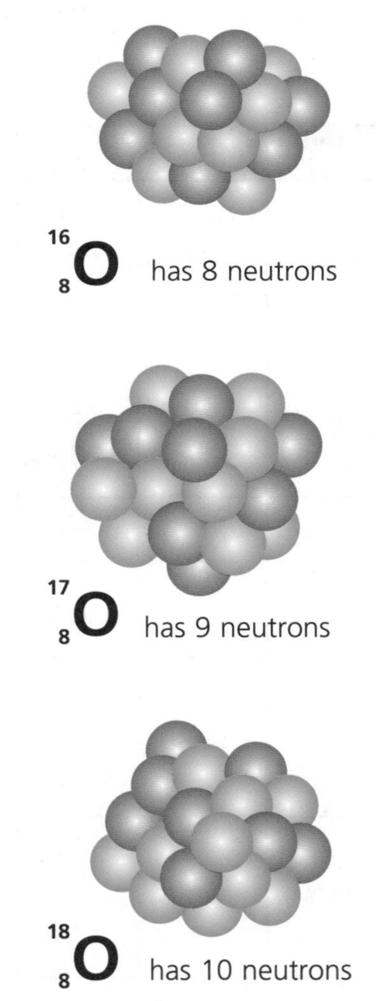

Background Radiation

Radiation occurs naturally all around us. This is known as background radiation. It only provides a very small dose altogether so there is no danger to our health.

13% of radiation is from man-made sources

87% of radiation is from natural sources

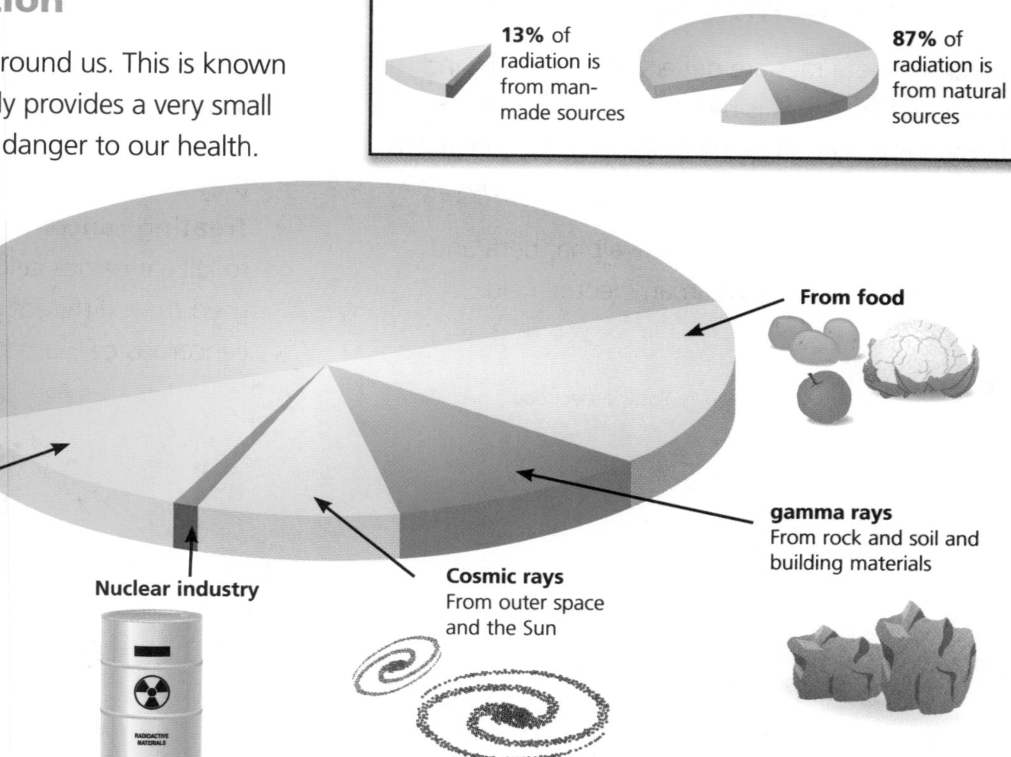

Radiation

Some substances give out radiation from the nuclei of their atoms all the time, regardless of what is done to them. These substances are said to be **radioactive**. Radiation is released from the nucleus as the result of a change in the structure of the atom.

When radiation collides with atoms or molecules, it can knock electrons out of their structure, creating a charged particle called an **ion**.

There are three types of radiation:

- **Alpha (α)** – an alpha particle is a helium nucleus (a particle made up of two protons and two neutrons).
- **Beta (β)** – a beta particle is a high-energy electron that is ejected from the nucleus.
- **Gamma (γ)** – high-frequency electromagnetic radiation.

The relative ionising power of each type of radiation is different, as is their power to penetrate different materials and their range in air.

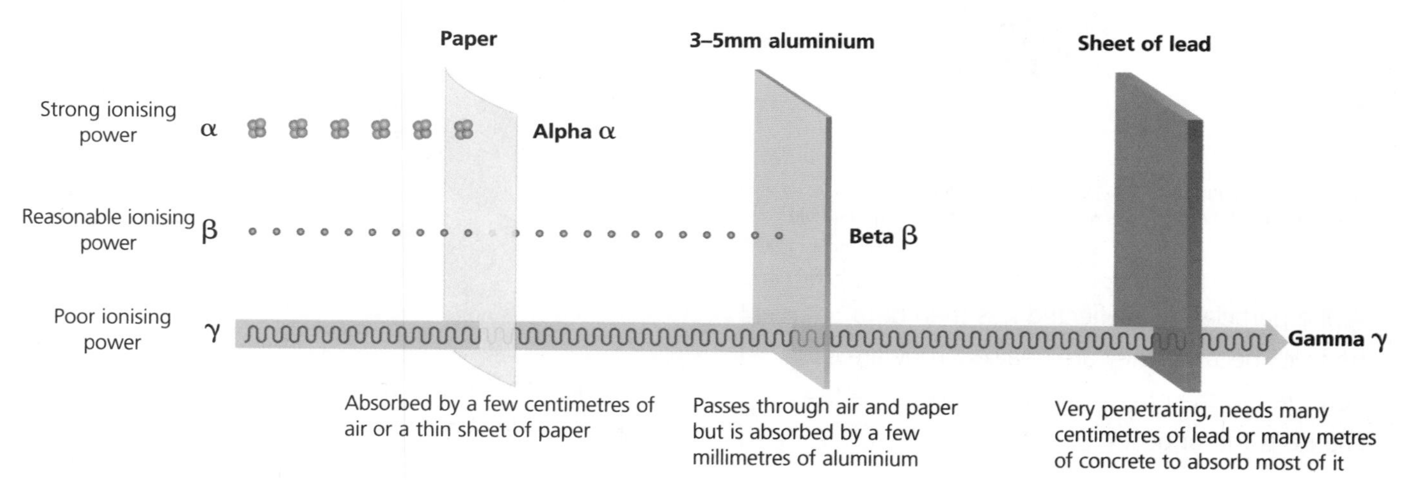

Electric and Magnetic Fields

Alpha and beta rays are deflected by electric and magnetic fields. This is because they are composed of charged particles. Gamma radiation is not deflected because it is not made up of charged particles.

The diagram below shows how alpha, beta and gamma radiation behave in an electric field.

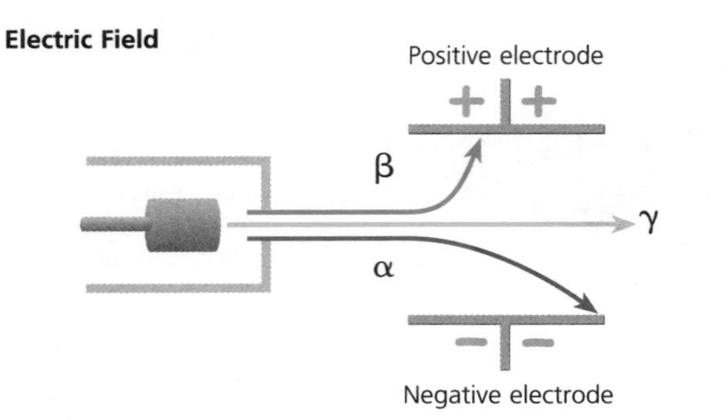

- **Alpha** (α)(→) particles are positively charged and are deflected towards the negative electrode.
- **Beta** (β)(→) particles are negatively charged and are deflected towards the positive electrode.
- **Gamma** (γ)(→) radiation is not deflected by the electric field.

The diagram below shows how alpha, beta and gamma radiation behave in a magnetic field.

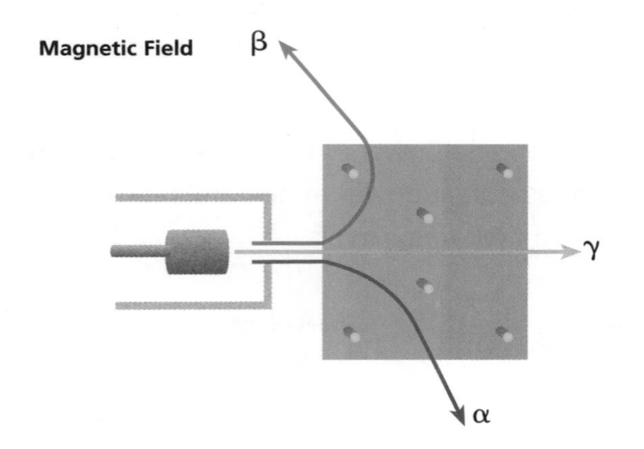

- **Alpha** (→) and **beta** (→) particles are deflected.
- **Gamma** (→) radiation is not affected by the magnetic field.

HT Alpha particles are deflected less than beta particles because they are heavier. They are deflected in the opposite direction because their charge is the opposite of beta particles.

Common Uses of Radiation

There are many uses for radiation:

- **Sterilisation** – gamma rays can be used to sterilise medical instruments and food because they kill bacteria.
- **Treating cancer** – gamma radiation can be used to kill cancerous cells. A high calculated dose is used from different angles so that only the cancerous cells are killed.

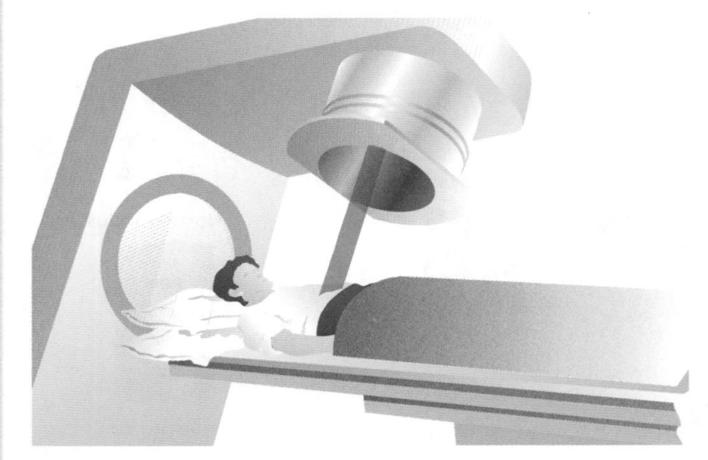

- **Tracers** – a tracer is a small amount of a radioisotope (radioactive isotope), which is put into a system. Its progress through the system can be traced using a radiation detector.
- **Controlling the thickness of materials** – when radiation passes through a material some of it is absorbed. The greater the thickness of the material, the greater the absorption of radiation. This can be used to control the thickness of some manufactured materials, e.g. paper production at a paper mill.

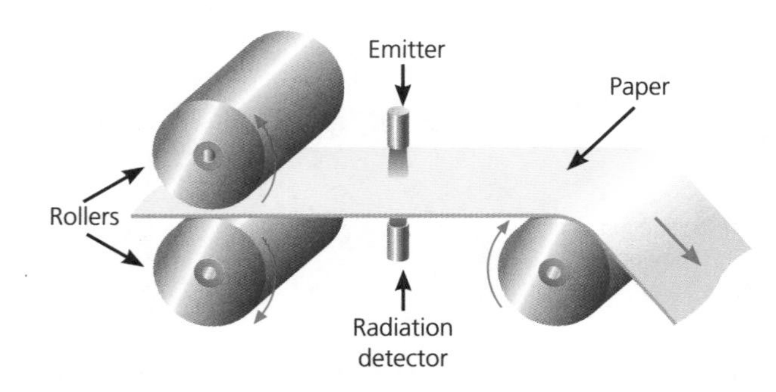

If the paper is too thick then less radiation passes through to the detector, and a signal is sent to the rollers which move closer together.

Acute Dangers

The ionising ability of alpha, beta and gamma radiation can damage molecules inside healthy living cells – this can result in the death of the cell.

Damage to cells in organs can cause cancer. The larger the dose of radiation the greater the risk of cancer.

The damaging effect of radiation depends on whether the source is located inside or outside the body. If the source is outside the body:

- alpha particles cannot penetrate the body and is stopped by the skin
- beta and gamma radiation can penetrate the body to reach the cells of organs and be absorbed by them. Gamma is less likely to be absorbed than beta.

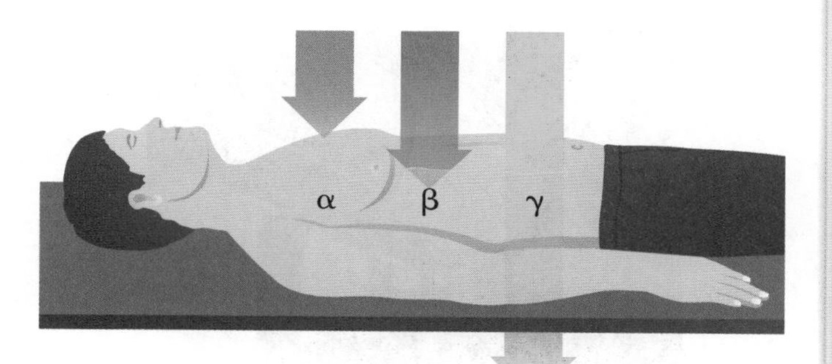

If the source is inside the body:

- alpha radiation causes most damage because it is easily absorbed by cells causing the most ionisation
- beta and gamma radiation cause less damage because they are less likely to be absorbed by cells.

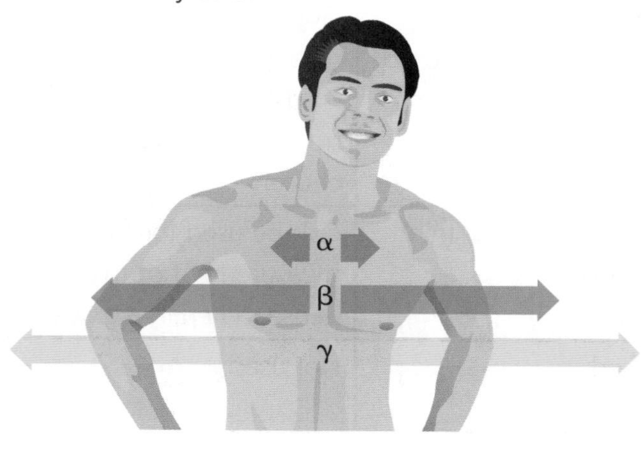

Half-life

The half-life of a radioactive isotope provides information about the rate of radioactive decay. It is the average time it takes for:

- The number of nuclei of the isotope (that have not decayed) in a sample to halve

  or
- The count rate (the number of atoms that decay in a certain time) of a sample containing the isotope to halve.

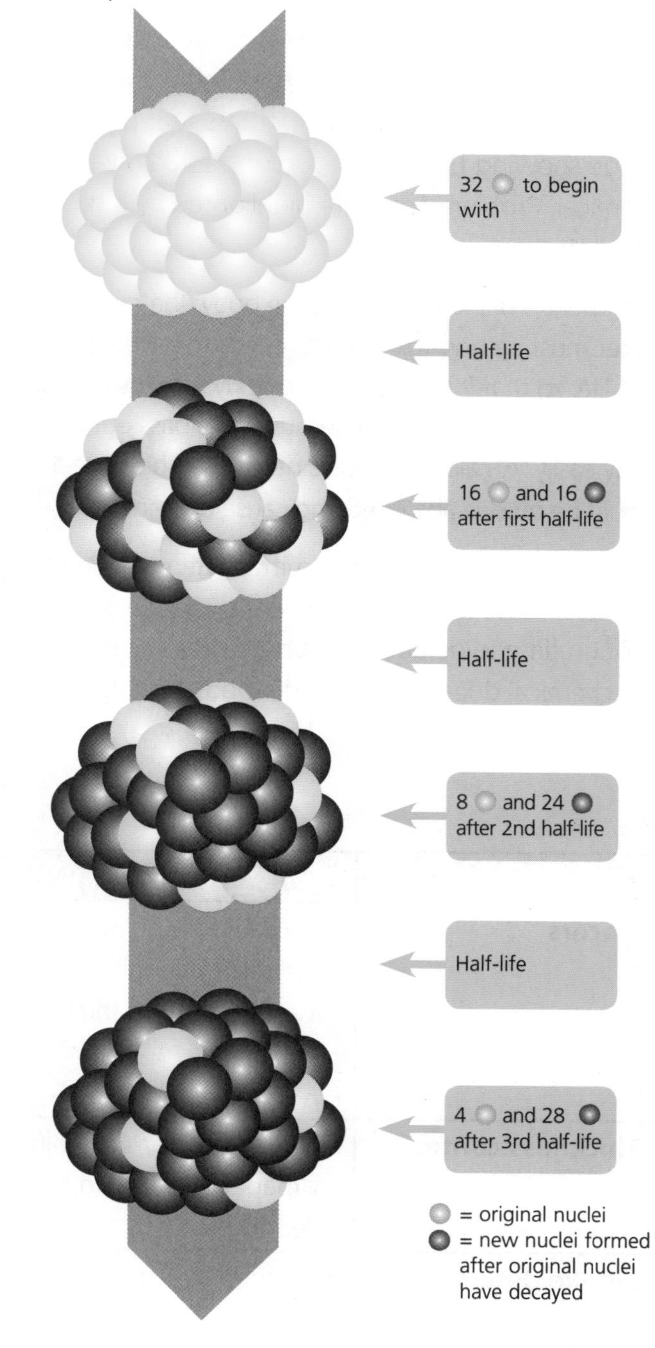

You need to be able to evaluate the appropriateness of radioactive sources for particular uses, including as tracers, in terms of the types of radiation emitted and their half-lives.

The properties of different radioactive substances make them suited to different uses.

Example 1 – Medical tracers
Doctors use radioactive chemicals called tracers to help detect damage to internal organs. Once the tracer enters the body, it builds up in the damaged or diseased part of the body. Radiation detectors can then be used to locate where the problem is. The detectors can be linked to computers that produce an image showing the distribution of the radioactive chemical. Problem areas are highlighted by a high concentration.

Because it is being used inside the body, the radioactive tracer must be non-toxic. It also needs to have a short half-life, so that it breaks down quickly after use. Gamma and beta sources are used because they pass out of the body easily. An alpha source is never used because it would be quickly absorbed and cause damage.

Technetium-99 is a gamma emitter often used for this purpose. It has a half-life of 6 hours. This gives doctors enough time to detect the problem, but ensures that the chemical does not stay in the body for too long.

Example 2 – Industrial tracers
Any leaks in a pipeline can be found by injecting a gamma-emitting isotope into the system. The tracer will leak out into the soil where the pipe is broken. Because of the penetrating nature of gamma rays, the leak is easy to detect through several feet of soil.

A short half-life radioisotope is used so that it does not remain in the environment any longer than is necessary after the leak has been detected.

Example 3 – Controlling the thickness of materials
Radiation is absorbed by material through which it passes. This effect can be used to monitor the thickness of materials in production. Alpha particles can be used to control the thickness of sheet paper, and beta particles can be used to control the thickness of thicker sheet paper and thin metal.

Use	Alpha	Beta	Gamma
Tracers	Not used because not penetrative enough to escape; harmful in the body	Not penetrative enough for industrial use but used for medical tracers; short half-life chosen	Good penetrative power means can be used for medical and industrial tracers; short half-life chosen
Thickness control	Can be used for ultrathin foil/paper that would not affect beta	Used for paper and thin metal foils	Can be used for thickness measurement of thick or dense materials e.g. lead
Smoke detectors	Absorbed by smoke particles	Not used	Not used

You need to be able to explain how the Rutherford and Marsden scattering experiment led to the 'plum pudding' model of the atom being replaced by the nuclear model.

In 1897, J.J. Thomson, a physicist, reasoned that since electrons were responsible for only a very small proportion of an atom's mass, they would take up an equally small proportion of an atom's size. He proposed that an atom consisted of a positive sphere of matter in which negative electrons were embedded.

The resulting model looked rather like a plum pudding, so it was therefore known as the 'plum pudding' model of the atom.

However, this model was disproved in 1911 by Ernest Rutherford, a British physicist, in his gold foil alpha scattering experiment.

Geiger and Marsden, working with Rutherford, placed a thin piece of gold foil in the centre of a circular chamber lined with a zinc sulfide detector. A radioactive source emitting alpha particles was focused on the gold foil target and the results were observed through a microscope.

Most alpha particles were seen to pass straight through the gold foil; this indicated that the gold atoms were composed of large amounts of open space. However, some particles were deflected slightly and a few were deflected back towards the source. This indicated that the alpha particles passed close to something positively charged within the atom and were repelled by it.

These observations brought Rutherford to conclude three things:

1. Gold atoms, and therefore all atoms, consist largely of empty space with a small, dense positive core. He called this core the **nucleus**.
2. The nucleus is positively charged.
3. The electrons are arranged around the nucleus with a great deal of space between them.

This is called the **nuclear atomic model**.

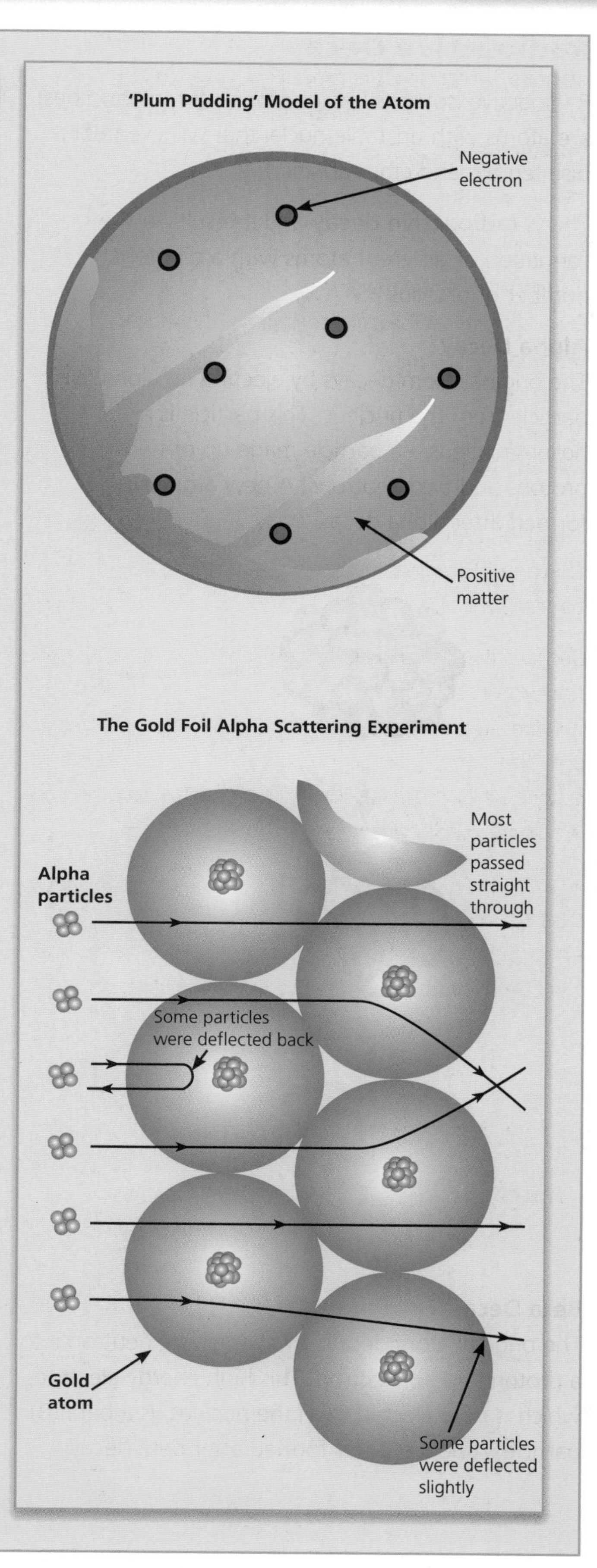

The Decay of Radioactive Substances

Radioactive Decay

Radioactive isotopes (radioisotopes or radionuclides) are atoms with unstable nuclei that will eventually disintegrate and emit radiation.

This is **radioactive decay** and it results in the formation of different atoms with a different number of protons.

Alpha Decay

The original atom decays by ejecting an alpha (α) particle from the nucleus. This particle is a helium nucleus – a particle made up of two protons and two neutrons. A new atom is formed after alpha decay.

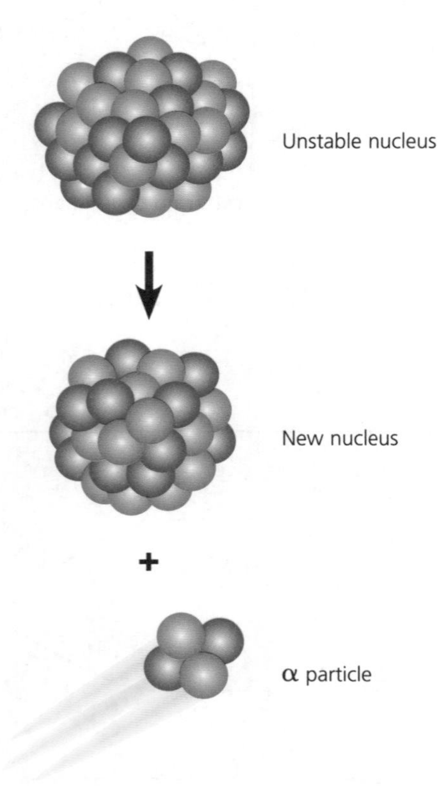

Beta Decay

The original atom decays by changing a neutron into a proton and an electron. This high energy electron, which is now ejected from the nucleus, is a beta (β) particle. A new atom is formed after beta decay.

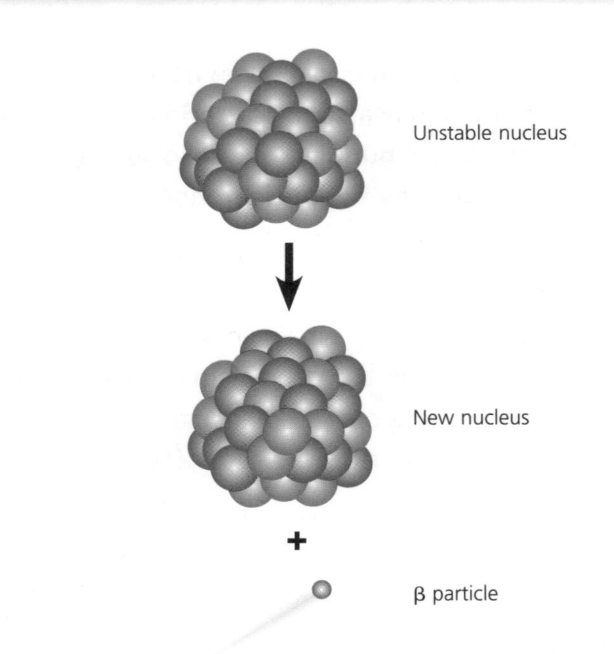

Gamma Decay

There is another type of radiation – gamma (γ) radiation. However, unlike alpha and beta, gamma emissions have no effect on the structure of the nucleus.

HT Decay Equations

You need to be able to write balanced decay equations for these nuclear changes by completing the atomic or mass numbers.

In alpha decay the mass number drops by 4 and the atomic number drops by 2.

$$^{241}_{95}\text{Am} \longrightarrow {}^{237}_{93}\text{Np} + {}^{4}_{2}\alpha$$

In beta decay the mass number is unchanged and the proton number goes up by one.

$$^{14}_{6}\text{C} \longrightarrow {}^{14}_{7}\text{N} + {}^{0}_{-1}\beta$$

In both cases the atomic number and mass number on the left-hand side are equal to the sum of the atomic numbers and mass numbers respectively on the right-hand side

P2.6 Nuclear fission and nuclear fusion

Nuclear fission is the splitting of atomic nuclei; nuclear fusion is the joining together of atomic nuclei. To understand this, you need to know:

- how uranium-235 and plutonium-239 are used
- how a chain reaction can be created
- how energy is released in stars.

Nuclear Fusion

Nuclear **fusion** is the **joining** together of two or more atomic nuclei to form a larger atomic nucleus. It takes a very high temperature – over 100 million degrees Celsius – to force the nuclei to fuse, like in the Sun.

Nuclear fusion generally releases more energy than it uses, which makes it self-sustaining, i.e. some of the energy produced is used to drive further fusion reactions. This is how stars release energy. In the core of the Sun, hydrogen is converted to helium by fusion. This provides the energy to keep the Sun burning and allow life on Earth.

The fusion of two heavy forms of hydrogen (deuterium and tritium) is an example of nuclear fusion.

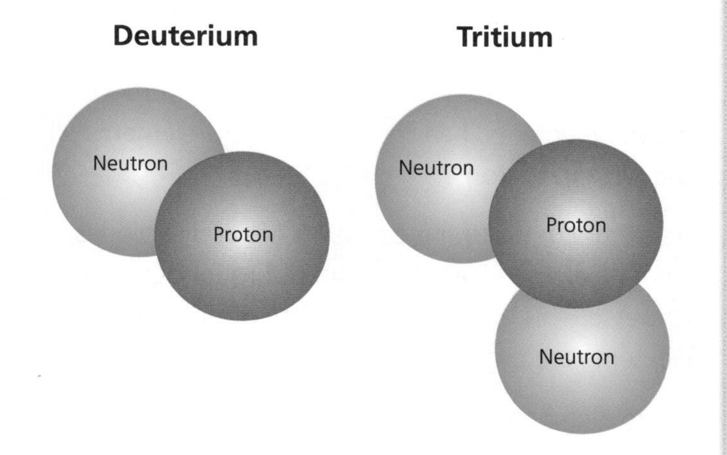

When they are forced together, the deuterium and tritium nuclei fuse together to form a new helium atom and an unchanged neutron.

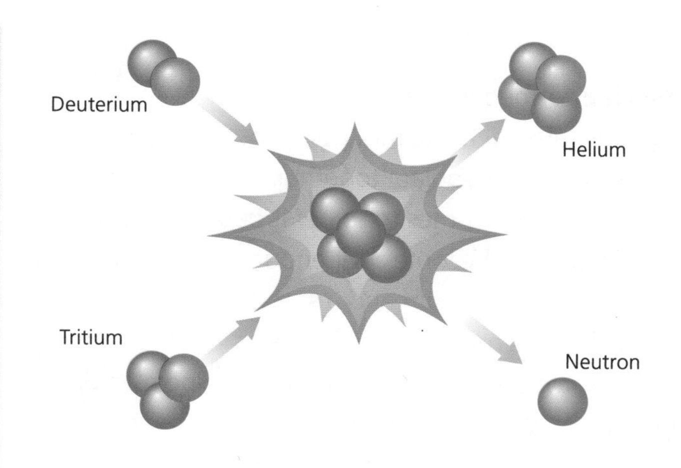

Nuclear Fission

Nuclear fission is the process of splitting large atomic nuclei. It is used in nuclear reactors to produce energy to make electricity. The two substances commonly used are uranium-235 and plutonium-239. For fission to occur the nucleus must first absorb a neutron.

When fission occurs the nucleus splits into two smaller nuclei emitting two or three neutrons and releasing energy. The neutrons released can be absorbed by another uranium nucleus and cause a chain reaction.

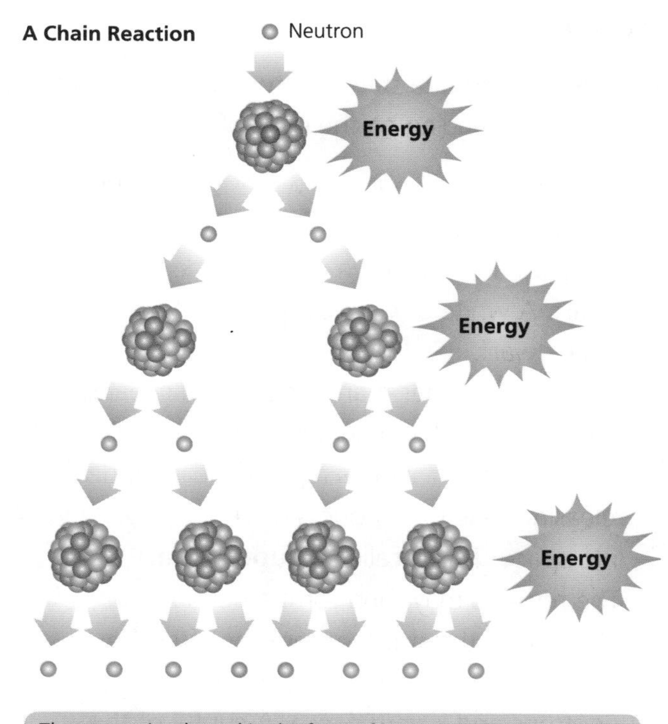

The energy is released in the form of heat. Each fission reaction produces only a tiny amount of energy, but there are billions of reactions every second.

Formation of Stars

Stars are formed in the following way:

1. Gravitational attraction pulls clouds of gas and dust (called nebulae) together to form a protostar.
2. As this mass comes together it becomes hotter. Eventually it becomes hot enough for hydrogen to fuse to form helium and a star is formed. This nuclear fusion releases massive amounts of energy and produces all the naturally occurring elements.
3. Dust may also clump together to form larger masses and eventually planets.

Stars use hydrogen as their energy source, which means they can release energy for billions of years. Our Sun is believed to be 5 billion years old and only halfway through its life.

The Life Cycle of a Star

A star remains stable during its life due to the balance of forces within it – mainly the radiation pressure of fusion creating an explosive outwards force and the force of gravity acting inwards. While these forces are balanced a star is called a **main sequence star**.

Towards the end of a star's life, two different processes may occur depending on the mass of the star:

- A star the size of our Sun will expand to become a red giant. The red giant then cools down and collapses under its own gravity to become a white dwarf. A white dwarf slowly cools to become a black dwarf.
- Stars much bigger than our Sun can expand enormously to become red supergiants. The red supergiant then shrinks rapidly and explodes, releasing massive amounts of energy, dust and gas into space. This is called a **supernova**.
  The remains may then form a neutron star. This is the core of the star that remains after the explosion. A neutron star is made only of neutrons and is very dense. A cupful of this matter could have a mass greater than 15 000 million tonnes!

The largest neutron stars collapse further forming black holes. They are so dense that nothing can escape from their very strong gravitational field – not even light.

Black holes can only be observed indirectly through their effects on their surroundings, e.g. the X-rays emitted when gases from a nearby star spiral into a black hole.

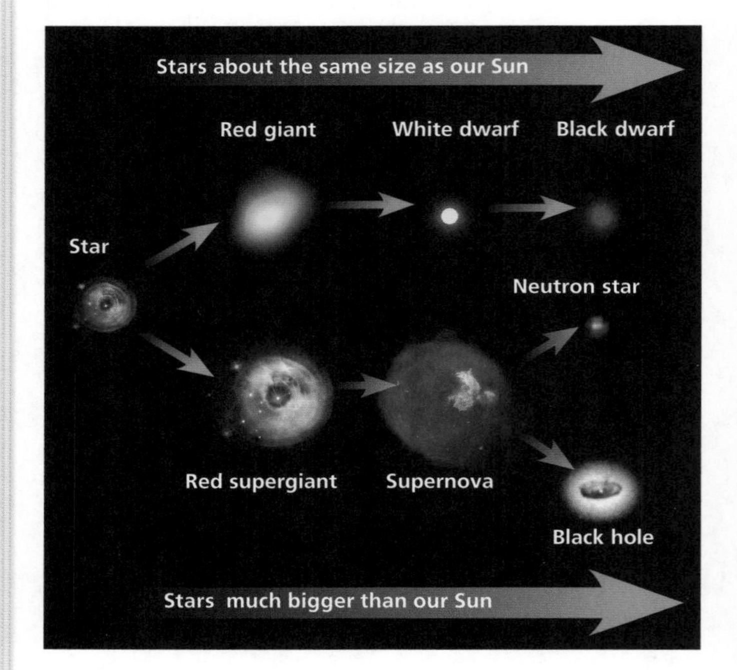

Recycling Stellar Material

Stars use hydrogen as a fuel to undergo nuclear fusion, which produces helium. Hydrogen and helium can also fuse together to produce nuclei of heavier elements; heavier stars can fuse elements all the way up to iron.

As a star comes to the end of its life and explodes, all of its elements are distributed throughout the Universe. This means that a variety of different elements are circulated in the Universe.

These elements can be recycled in the formation of new stars or planets. Atoms of heavier elements are present in the inner planets of our **Solar System**, which leads us to believe that our Solar System was formed from the material produced when earlier stars exploded.

1 The diagram shows a fusion reaction:

a) What happens during a fusion reaction? **(2 marks)**

b) i) Which two of the labelled nuclei are isotopes of the same element? **(1 mark)**

ii) How do you know? **(2 marks)**

c) What is released during the fusion reaction? **(1 mark)**

d) A fusion reaction takes place in a star.

i) What force balances the radiation force of fusion for most of a star's life? **(1 mark)**

ii) Towards the end of its life this force gets weaker and the star expands and cools. What is this stage of a star's life called? **(1 mark)**

iii) Describe the rest of the star's life cycle for the largest stars. **(3 marks)**

2 The diagram shows the forces acting on a van of mass 3000kg.

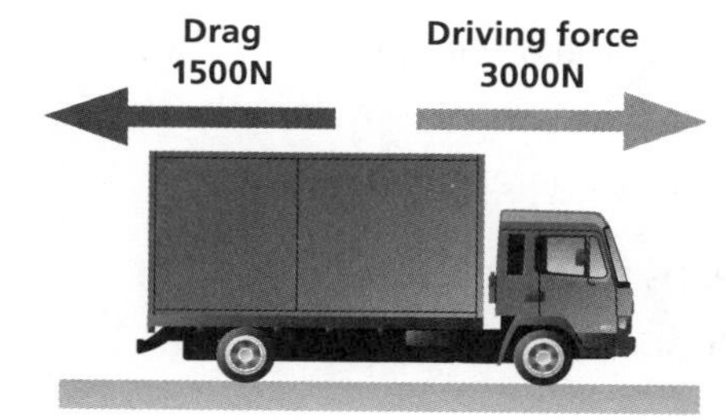

a) Calculate the acceleration of the van. **(4 marks)**

b) After a while, the van stops accelerating and reaches a top speed, even though the driving force has not changed. Explain, by referring to the forces acting, why this happens. **(3 marks)**

c) A new design of van with the same mass and engine has a higher top speed than the old style pictured. Suggest why this happens. **(1 mark)**

3 The text below shows the information for two different vacuum cleaners for sale in a shop.

Model X:
230V, 1000W power, quite lightweight, £120

Model Y:
230V, 1500W high power, £150

a) What is power a measure of? **(1 mark)**

b) When working at maximum power, calculate the current supplied to model X. Show your working. **(3 marks)**

c) Fuses are available as 3A, 5A and 13A. Which fuse should be used for model X? **(1 mark)**

d) Model Y is more powerful but many customers still choose to purchase model X. Suggest two reasons why. **(2 marks)**

Medical Applications of Physics

P3.1 Medical applications of Physics

There are a number of different medical applications of Physics. To understand this, you need to know:

- the use of X-rays for diagnosis and treatment
- the use of ultrasound for diagnosis and treatment
- the use of lenses for the correction of defects of vision
- that the range of human hearing is about 20Hz–20 000Hz
- the structure of the eye and how we correct defects in vision.

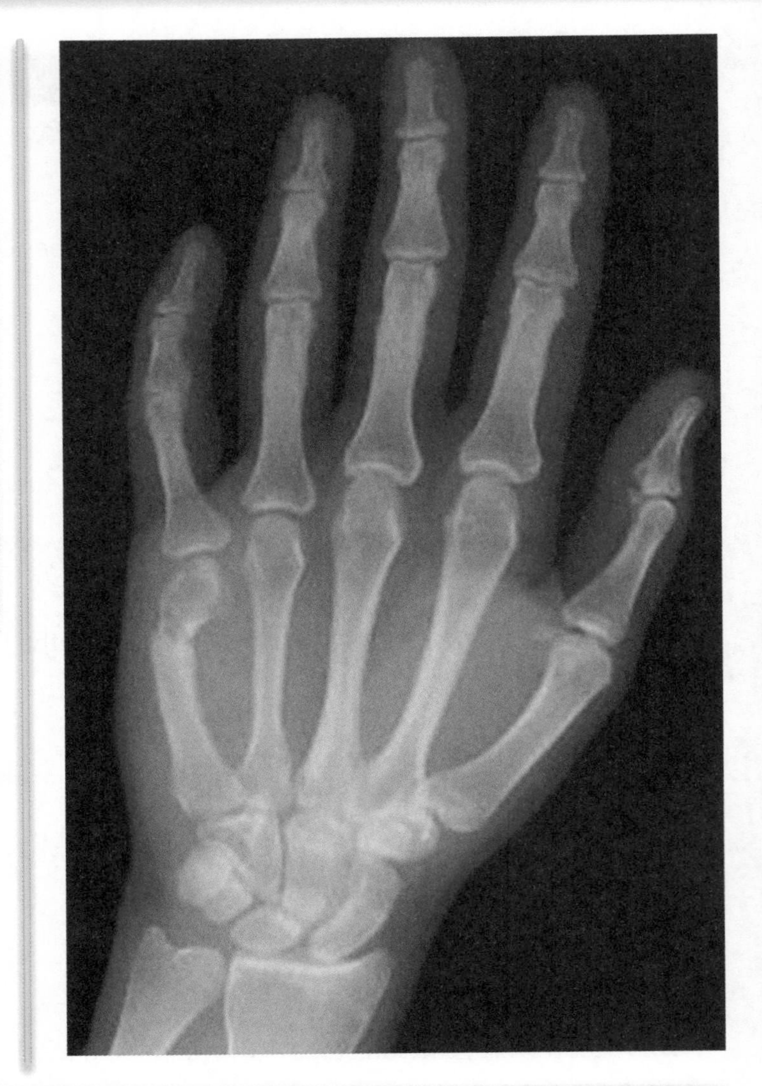

X-Rays

X-rays are part of the electromagnetic spectrum. They have a high frequency and a very short wavelength (about the same size as the diameter of an atom).

The high frequency means that they have a high energy and can cause ionisation.

As with all uses of waves, the properties of X-rays are linked to their use and these are detailed in the table below.

Property	Use	Precautions
Affect photographic film (in the same way that light does)	Can form an image on an X-ray plate or with a charge coupled device (CCD). CCDs allow the image to be formed electronically	All uses of X-rays have the same precautions: • Only use if necessary and use as low a power as possible • Use a focused beam during treatment so as not to expose areas of the body not being treated or examined • Technicians should take the X-rays from behind a screen or wear lead clothing • Technicians should wear detection badges, which measure their exposure over time.
Absorbed by dense materials (bone and metals)	Can take shadow pictures of bones to detect breaks and fractures	
Transmitted by healthy tissue	Used for computerised tomography (CT) scans to detect problems with soft tissue, e.g. disease, tumours, embolisms and haemorrhages	
High ionising power	Used for killing cancer cells in radiotherapy	

Ultrasound

Ultrasound is sound waves of frequencies greater than 20 000Hz, i.e. above the upper limit of the hearing range for humans. Electronic systems produce electrical oscillations, which are used to generate ultrasonic waves.

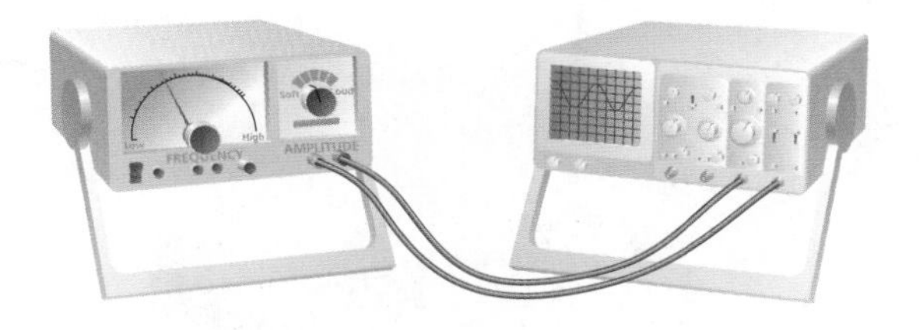

As ultrasonic waves pass from one medium (substance) into another they are partly reflected at the boundary. The time taken for these reflections is a measure of how far away the boundary is. The reflected waves are usually processed to produce a visual image on a screen.

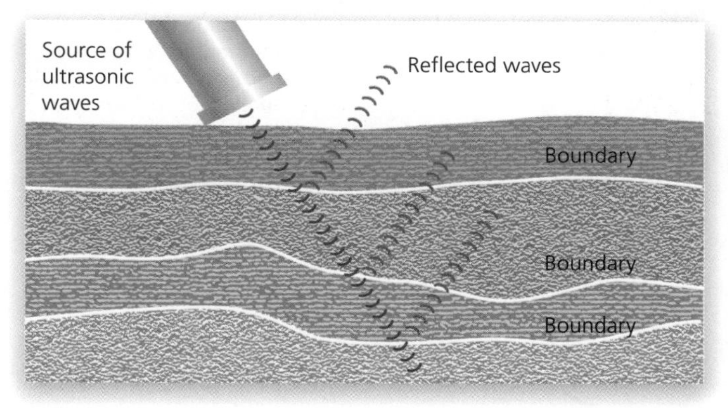

Uses of Ultrasound

Use in medicine

The main use of ultrasound in medicine is in pre-natal scans, but it can also be used for other scans, e.g. detection of testicular cancer.

An ultrasound pulse is sent into the body and get reflected back at different changes of density. These reflected waves are detected and used to form a picture of the foetus. Because ultrasound is non-ionising it is safer than using X-rays.

Ultrasound can also be used to break up kidney stones, which can then be passed out of the body in urine. This saves having to operate on the patients' kidneys.

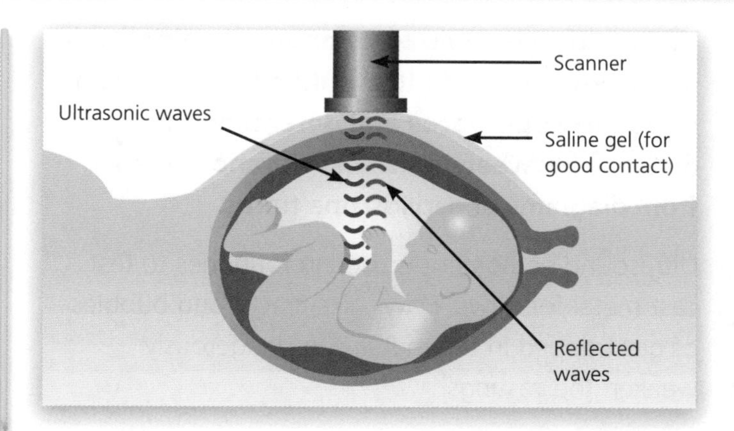

Detecting flaws and cracks

Some of the ultrasound waves are reflected back by the flaw or crack within the structure. The time taken for the reflected wave to return is used to calculate the location of the crack.

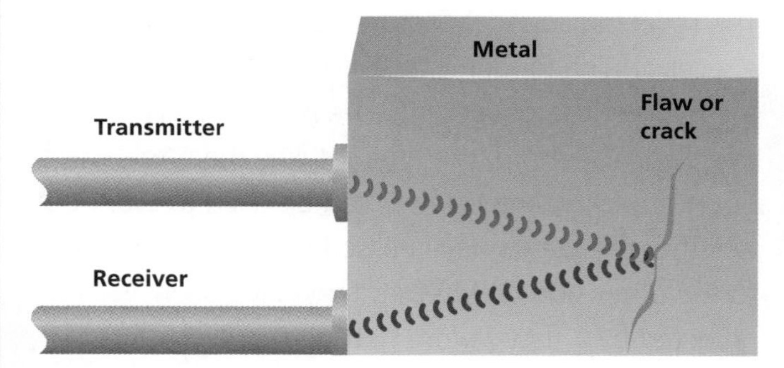

Cleaning delicate objects

The vibrations caused by the ultrasonic waves can be used in a liquid to dislodge dirt particles from the surface of an object. Using this method means there is no danger of breakage and no need to take the object apart.

Comparing Scanning Methods

Method	Image Quality	Radiation Exposure
CT scans	Very good quality image	Longest exposure to ionising radiation
X-rays	Good quality image	Short exposure to ionising radiation
Ultrasound	Average to excellent depending on equipment	No exposure to ionising radiation

You need to be able to compare the amplitudes and frequencies of ultrasounds from diagrams of oscilloscope traces and determine the distance between interfaces in various media from diagrams of oscilloscope traces.

Ultrasound equipment is used in foundries to test cast metal for flaws. Flaws, like cracks and bubbles of gas trapped in the metal, can dangerously weaken the castings.

Ultrasound waves are transmitted through the castings. When they meet a boundary between substances of different densities – e.g. the other side of the metal object or a flaw in the metal – some of the waves are reflected back.

The reflected sound waves (echoes) are detected by a receiver. The longer it takes for the wave to reach the receiver the deeper into the metal it has travelled. Therefore, the time it takes them to bounce back to the receiver can be used to work out whether the waves have passed through the object to the other side or whether they have been reflected by a flaw before getting there.

Example 1

We want to calculate the depth of a flaw from the traces opposite. Each square on the traces represents 1 microsecond (1×10^{-6} second) and sound waves travel through the metal at 5000m/s.

Using the equation, Distance = Speed × Time

Distance = $5000\text{m/s} \times 3 \times 10^{-6}\text{s}$

(or 5000m/s × 0.000 003s)

= 1.5×10^{-2}m (or 0.015m)

= **1.5cm**

Remember, this is the distance there and back, so the depth of the flaw in the metal is actually half this amount.

$\frac{1.5}{2}$ = **0.75cm**

The flaw is 0.75cm beneath the surface of the metal in the casting.

Example 2

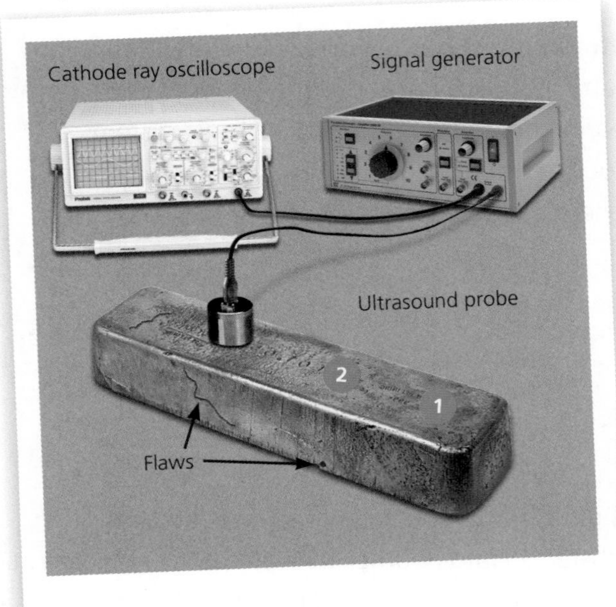

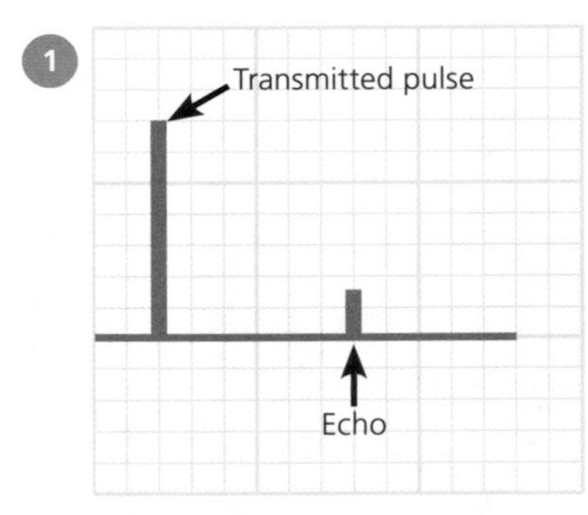

The echo has lower amplitude than the transmitted pulse, because waves spread out as they get further away from the source (i.e. amplitude decreases)

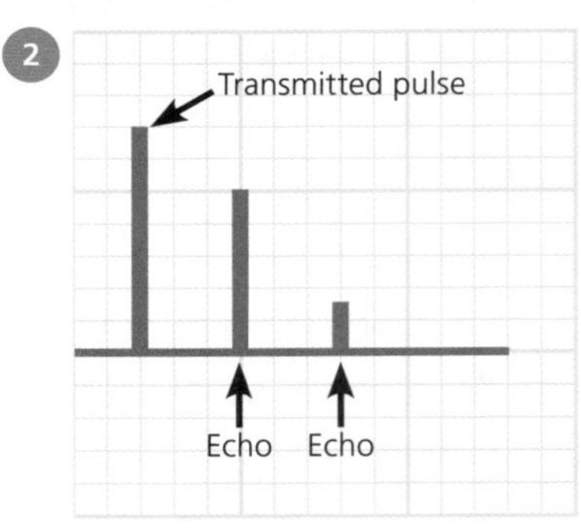

Above are traces produced at positions 1 and 2. Trace 1 shows an echo that has travelled cleanly through the metal and been reflected at the boundary on the other side. On Trace 2, as well as the echo from the end, an earlier echo is returned. This must have been reflected by a flaw in the metal.

Lenses

Refraction is the change of direction of light as it passes from one medium to another. The change of direction depends on the refractive index of the material and the angle of incidence.

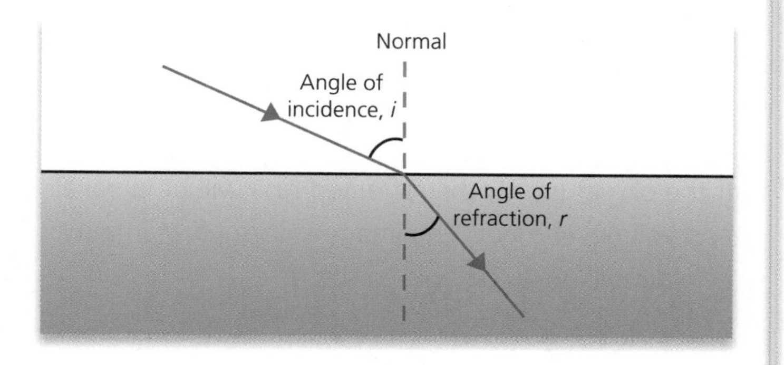

The diagram shows what happens during refraction. The refractive index of the blue material is higher than air, so the ray bends towards the normal.

If we measure the angles of incidence and refraction, we can calculate the refractive index of the material with the formula:

$$\textbf{Refractive index} = \frac{\sin i}{\sin r}$$

It is this bending of light that is used in lenses.

Images Produced by Lenses

A lens is a piece of transparent material that refracts light rays. There are two types of lens – **diverging** and **converging**. These lenses have a different curvature, therefore parallel rays of light pass through them differently.

For both types of lens, the distance from the lens to the principal focus is called the **focal length**. The magnification produced by the lens can be found with the formula:

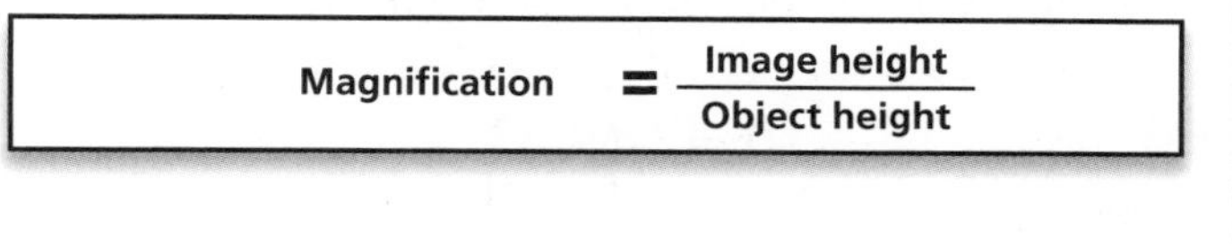

$$\textbf{Magnification} = \frac{\textbf{Image height}}{\textbf{Object height}}$$

Diverging Lens

A diverging (concave) lens is thinnest at its centre.

In a double concave lens, the rays of light are refracted outwards at the two curved boundaries so that they appear to come from one point, the focus (F). Only the middle ray, which enters the lens at 90°, passes straight through.

The image produced by a diverging lens is **virtual** and **upright**.

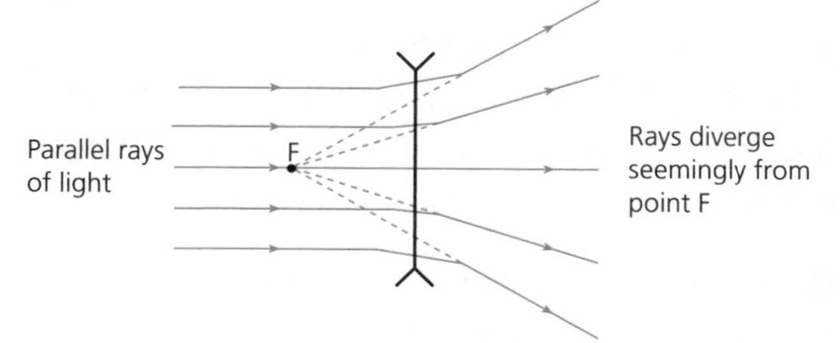

Converging Lens

A converging (convex) lens is thickest at its centre.

In a double convex lens, the light rays are refracted inwards at the two curved boundaries to converge (meet) at one point called the focus (F). Only the middle ray, which enters the lens at 90°, passes straight through.

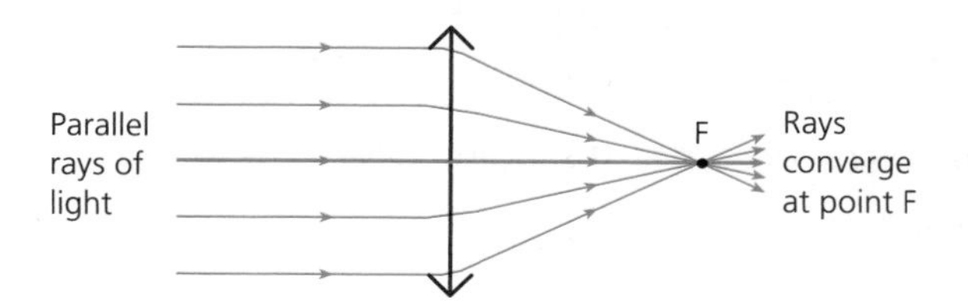

The image produced by a converging lens depends on the distance of the object from the lens:

- If the distance from the object to the lens is longer than the distance from the lens to the focal point (F), it will produce a **real image** that is **inverted**. This type of lens is used in cameras.
- If the distance from the object to the lens is **less** than the distance from the lens to the focus point (F)**,** it will produce a **virtual image** that is **upright**, **enlarged** and seems to be formed **on the same side** of the lens. This type of lens is used in magnifying glasses.

You need to be able to explain the use of a converging lens as a magnifying glass and construct ray diagrams to show the formation of images by diverging and converging lenses.

Ray diagrams can determine the path of light from an object to our eyes. By drawing them we can find out the location, size, orientation and type of image formed. Regardless of where the object is, the principle for drawing ray diagrams is the same.

To draw a ray diagram, first draw a horizontal axis and a double convex or concave lens. The horizontal axis should run through the centre of the lens. Mark the focal point (F) on both sides of the lens. They should be the same distance from the centre of the lens.

Now mark up 2F, again on both sides of the lens. Each 2F should be the same distance from F, as F is from the lens. Now mark the position of the object (i.e. outside 2F, at 2F, at F, etc.) There are three rays of light that can now be drawn from a point on the object.

- **Ray 1** runs parallel to the horizontal axis until it hits the lens. It will then refract through F.
- **Ray 2** goes straight through the centre of the lens and emerges from the lens undeflected.
- **Ray 3** passes through F until it hits the lens. It will then refract and travel parallel to the horizontal axis.

The image occurs at the point where these rays intersect.

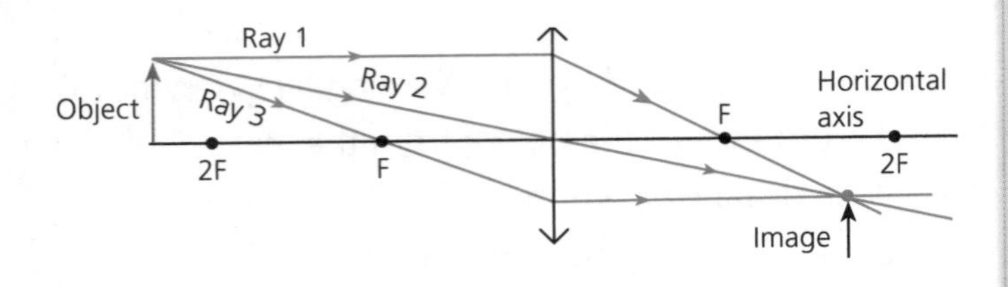

To find the location of an image you only need to draw two of these rays, so for the following diagrams we shall use rays 1 and 2.

Double Converging Lenses

The image produced with a converging lens depends on where the object is in relation to the focus point.

When the object is located **beyond 2F**, the image will be located on the other side of the lens. The image will be real, inverted and smaller. This type of lens is used in cameras.

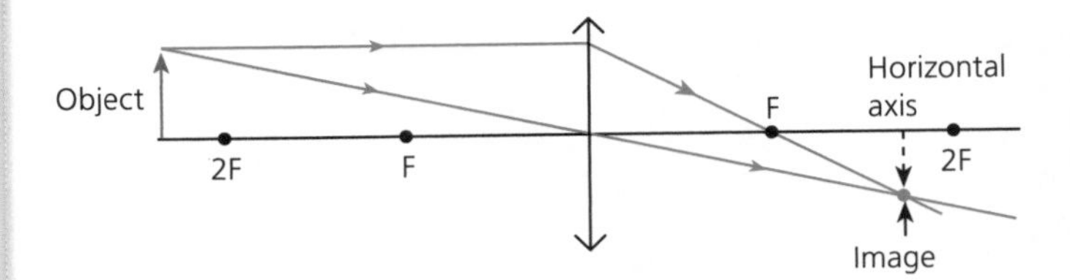

If the object is at F, the refracted rays will be parallel and will never meet. The image is formed at infinity, i.e. you will never be able to see it.

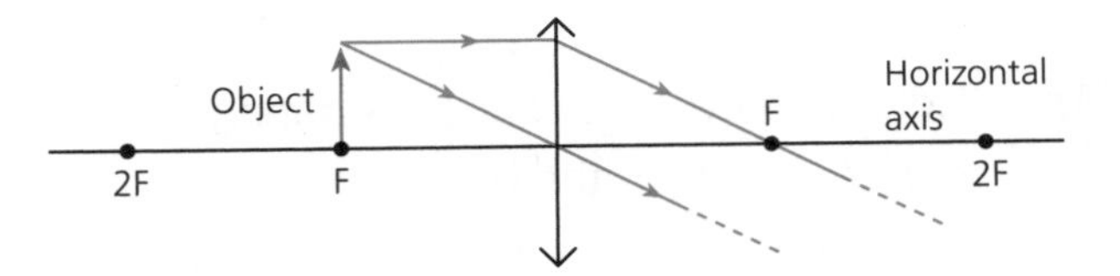

If the object is located in **front of F**, the light rays diverge after refracting through the lens. The image location can be found by tracing all light rays backwards until they intersect. This produces a virtual, upright and enlarged image. This is used in a magnifying glass.

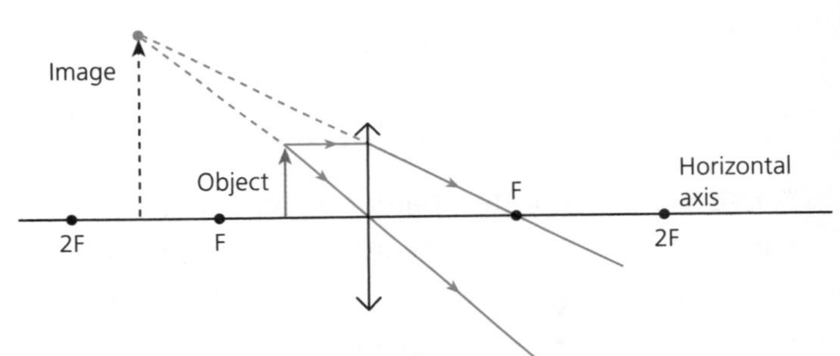

Double Diverging Lenses

The rays diverge after they are refracted. The image location can be found by tracing ray 1 backwards until it intersects ray 2. The image produced from a diverging lens will always be virtual and upright, regardless of the position of the object.

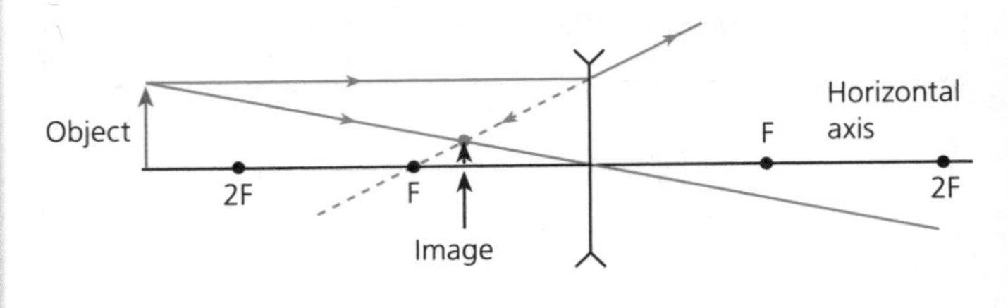

The Eye

The eye has an adjustable convex lens that is used to focus the image.

When a person needs glasses, it is because their lenses are unable to focus the image correctly on the retina.

A pair of glasses ensures that the image is focused on the retina.

The Structure of the Eye

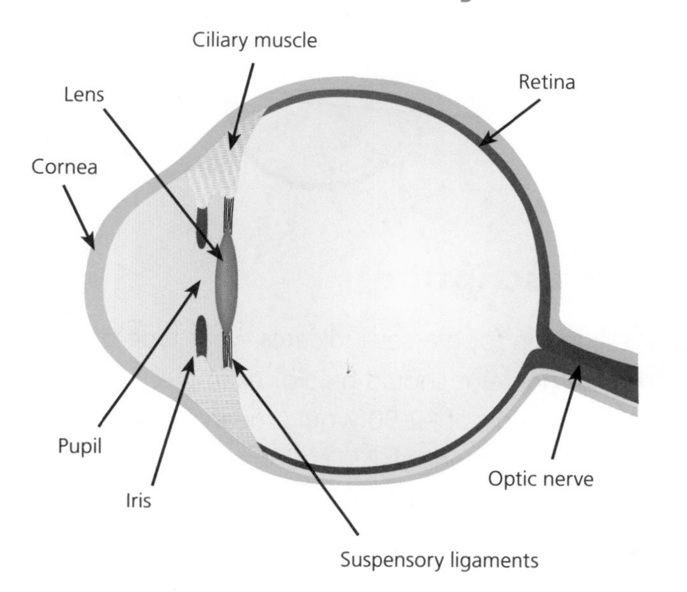

- **Retina** – images are focused here
- **Lens** – focuses the image
- **Cornea** – protects the eye and begins to focus the light
- **Pupil** – light enters the eye here
- **Iris** – adjusts to alter the amount of light entering the eye
- **Suspensory ligaments** – connect the lens to the ciliary muscles
- **Ciliary muscles** – contract and relax.

The ciliary muscles contract and relax to alter the shape of the lens so the eye can focus on objects at different distances.

Pulling on the suspensory ligaments pulls on the lens and makes it thinner, which allows it to focus on distant objects.

The Camera

The camera works in a similar way to the eye and many of the components of a camera mimic those of the eye.

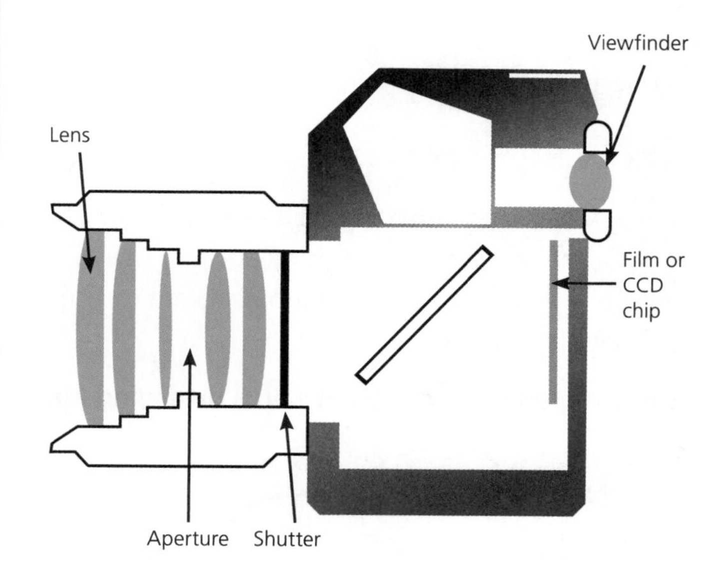

Comparing the Eye and a Camera

The table below shows how an eye and a camera can provide the same function.

Function	Eye	Camera
Allows light to enter	Pupil	Aperture
Focuses light	Cornea and lens	Convex lens
Adjusts focus for different distances	Muscles and ligaments alter shape of lens	Lens moves closer to or further away from the film / CCD chip
Controls amount of light entering	Iris changes pupil size	Diaphragm / aperture stop changes the aperture size
Forms the image	Retina	Photographic film or CCD chip (digital cameras)

Correcting Vision Defects

When the eye is working normally, the range of vision is between the near point (around 25cm) and the far point (infinity).

Objects closer than the near point cannot be brought into focus.

Short Sightedness

When a person is **short sighted**, the eyeball is too long or the lens is too fat. This means that near objects are in focus, but distant objects are focused in front of the retina.

Short sightedness is corrected with a **diverging lens**.

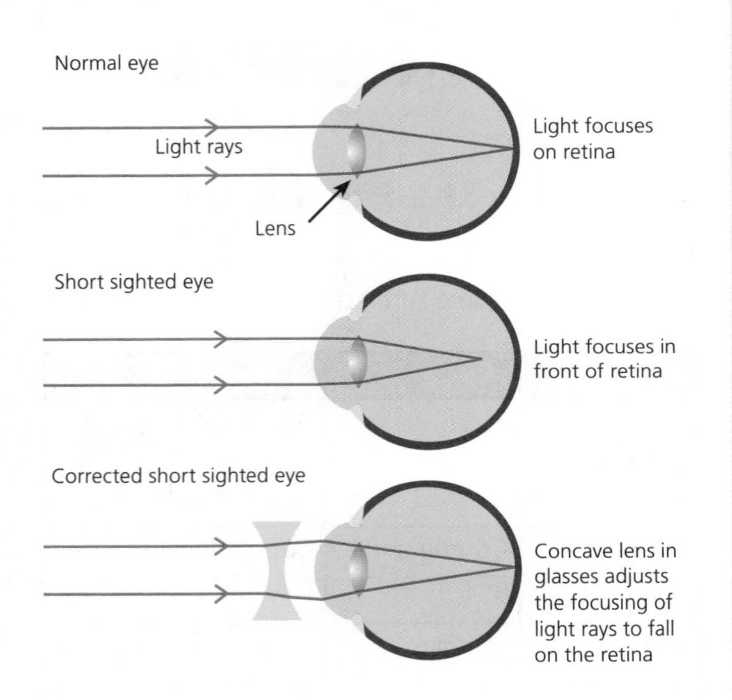

Long Sightedness

When a person is **long sighted**, the eyeball is too short or the lens is too thin. This means that distant objects are in focus, but near objects are focused behind the retina.

Long sightedness is corrected with a **converging lens**.

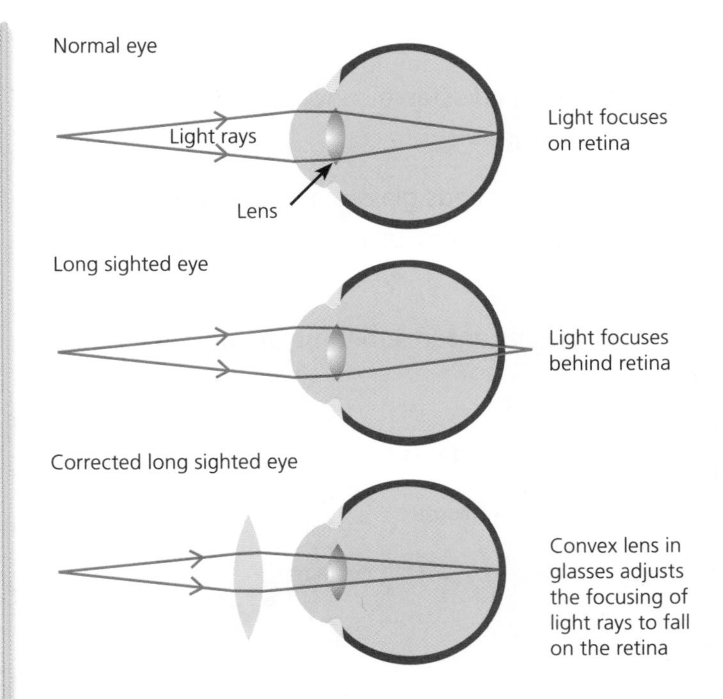

The Prescription

A prescription for glasses indicates what type of lens and what power it should be. For example:

- a prescription of –2.50 would mean a diverging lens with a power of 2.5D (dioptres)
- a prescription of +2.50 would mean a converging lens with a power of 2.5D.

The power of the lens is linked to the focal length by the equation:

$$\textbf{Power}\ (\text{D}) = \frac{1}{\textbf{Focal length}\ (\text{m})}$$

The focal length and, therefore, power of the lens depends on two things:

- The **curvature** – a more curved lens has a higher power and shorter focal length.
- The **refractive index** – a high refractive index has a higher power and shorter focal length.

HT People with a high prescription need more curved lenses. Thin and light lenses are made with a higher refractive index glass. This means the lens can be manufactured thinner while still having the correct focal length.

Total Internal Reflection

When a ray of light travels from glass, Perspex or water into air, some light is also reflected at the interface. This is called **internal reflection**.

Total internal reflection (where no light is refracted at all) occurs when the angle of incidence exceeds a certain value, called the critical angle. The **critical angle** for glass is approximately 42°.

Three possibilities are looked at below:

1. **Angle of Incidence < Critical Angle** – most light is refracted; there is a little internal reflection.

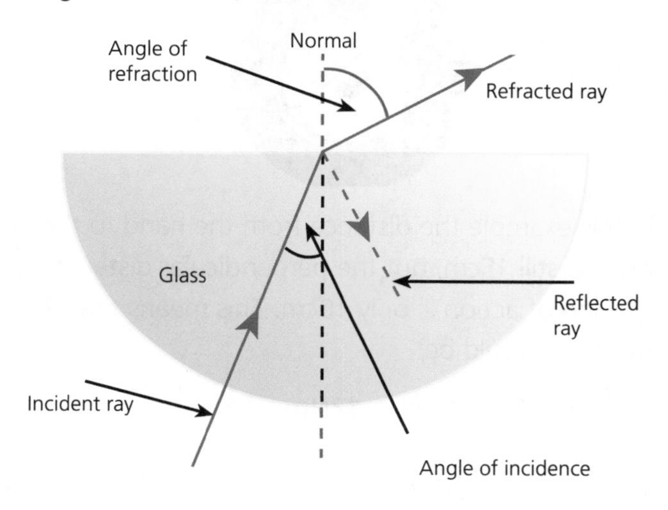

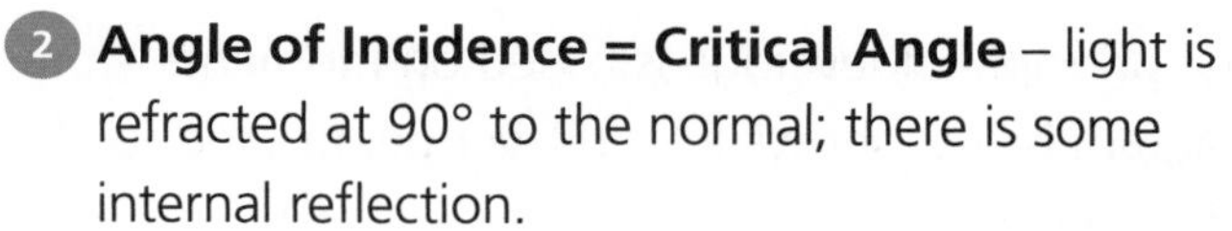

2. **Angle of Incidence = Critical Angle** – light is refracted at 90° to the normal; there is some internal reflection.

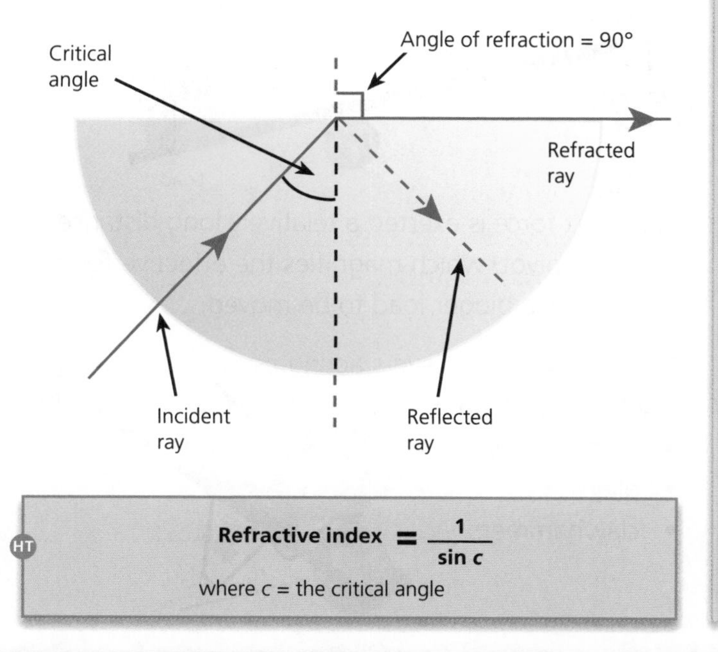

HT

Refractive index $= \dfrac{1}{\sin c}$

where c = the critical angle

3. **Angle of Incidence > Critical Angle** – total internal reflection; no refraction.

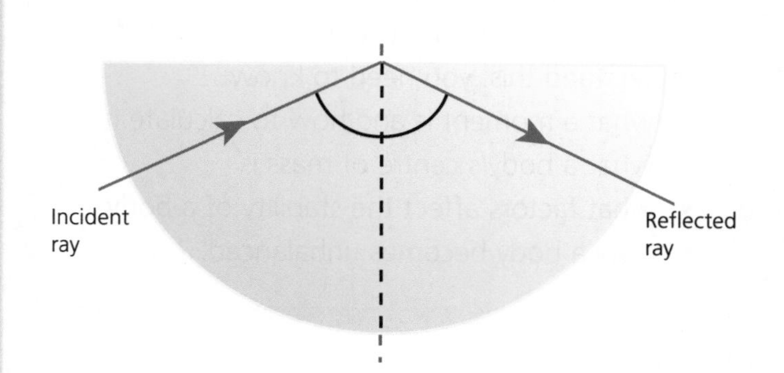

Total internal reflection is used to send light along optical fibres. This can be used for sending data for telephone calls and the Internet.

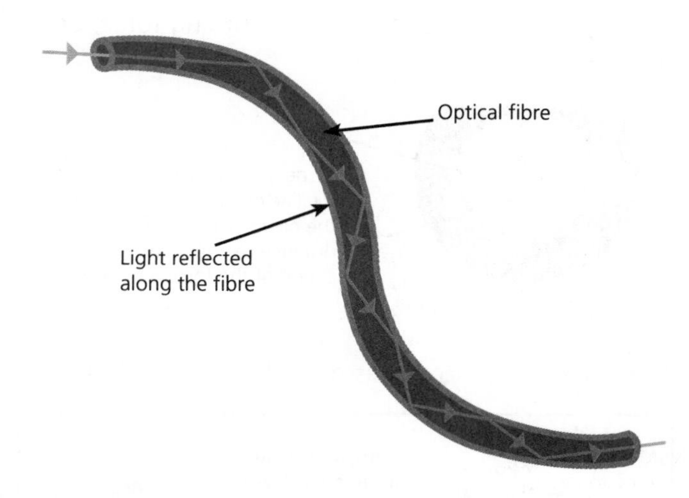

In medical applications, total internal reflection is used with an endoscope for internal examinations, which removes the need for surgery to see inside the patient's body. The optical fibre is used to provide illumination inside the body by shining a light through the fibre.

Lasers

Lasers have many uses in medical physics. The main use is as an energy source for cutting, burning and cauterising. This has many uses including:

- removing verrucas (by simply burning them)
- delicate laser eye surgery – the cornea is reshaped using a laser to enhance the bending of light that would normally be done by the lenses in a pair of glasses.

P3.2 Using physics to make things work

Balanced forces acting on a body can make the body turn but not change its speed. To understand this, you need to know:

- what a moment is and how to calculate it
- what a body's centre of mass is
- what factors affect the stability of a body
- how a body becomes unbalanced.

Moments

Forces can be used to turn objects about a pivot (fulcrum). The turning effect of a force is called the **moment**. If a spanner was used to unscrew a wheel nut, it would exert a moment, or turning force, on the nut.

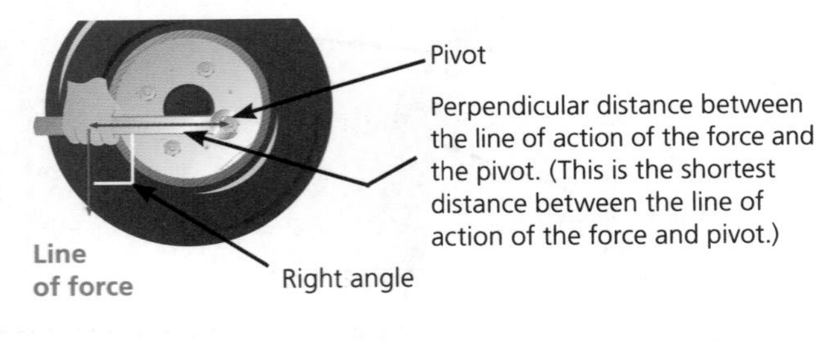

The size of the moment is given by the equation:

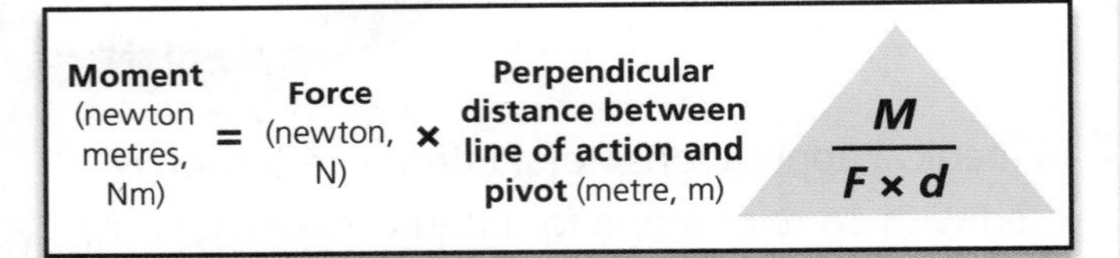

There are two ways of increasing the moment or turning force:

- increase the force applied
- increase the perpendicular distance between the line of action of the force and the pivot.

Example 1

A man unscrews a nut (pivot) on a wheel. The man exerts a force of 120N and the perpendicular distance from his hand to the pivot is 15cm. What moment does he exert?

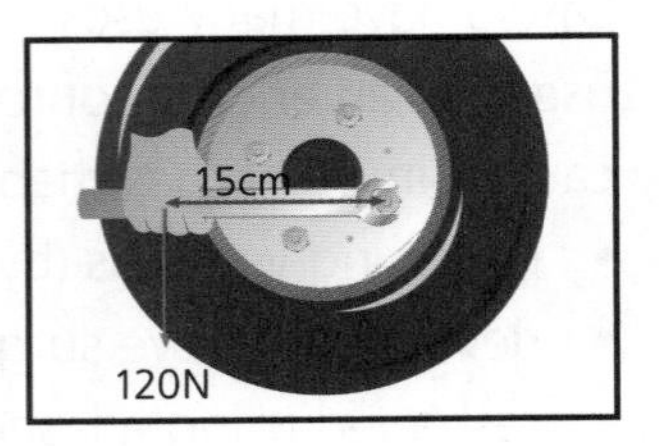

Moment = Force × Perpendicular distance between line of action and pivot

= 120N × 0.15m

= **18Nm**

The key point when doing these calculations is that it must be the *perpendicular* distance to the pivot that is used.

Example 2

In the example above, the line of action was at right angles to the spanner, but what happens when the line of action is not at a right angle?

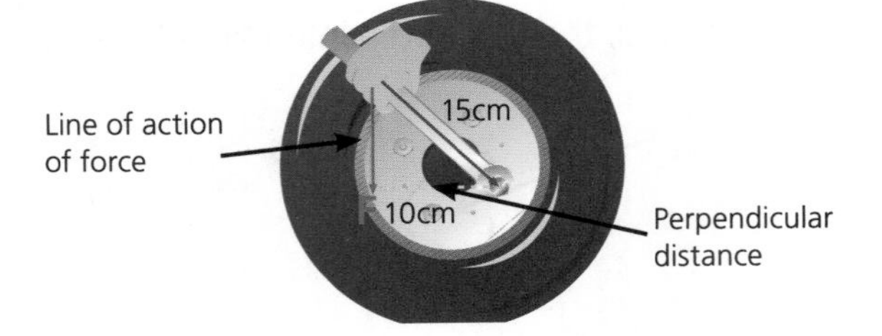

In this example the distance from the hand to the pivot is still 15cm, but the perpendicular distance to the line of action is only 10cm. This means that the moment would be:

F × d = 120N × 0.1m = **12Nm**

Simple Levers

The principle of leverage is based on moments – the diagram below shows a long lever being used to lift a heavy load.

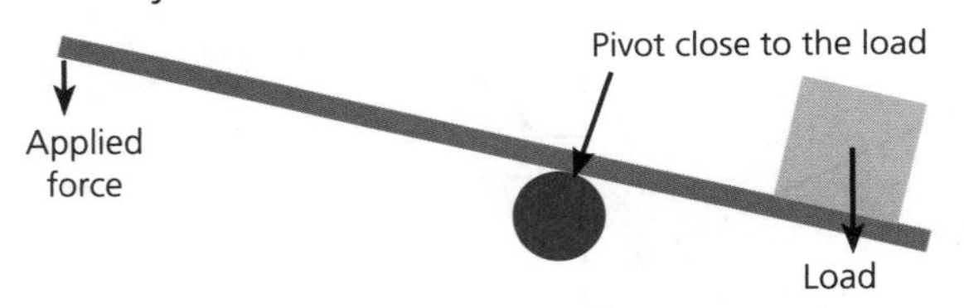

An effort force is exerted a relatively long distance from the pivot, which magnifies the effective force. This allows a bigger load to be moved.

Other examples of levers acting as force multipliers are:

- wheelbarrows
- pliers
- clawhammers.

Applied force
Load
Pivot

Centre of Mass

The **centre of mass** of an object is the point through which the whole weight (*W*) of the object is considered to act. If you were to balance an object on the end of your finger, the point at which the object balances is the centre of mass. The centre of mass of a symmetrical object is found along the axis of symmetry.

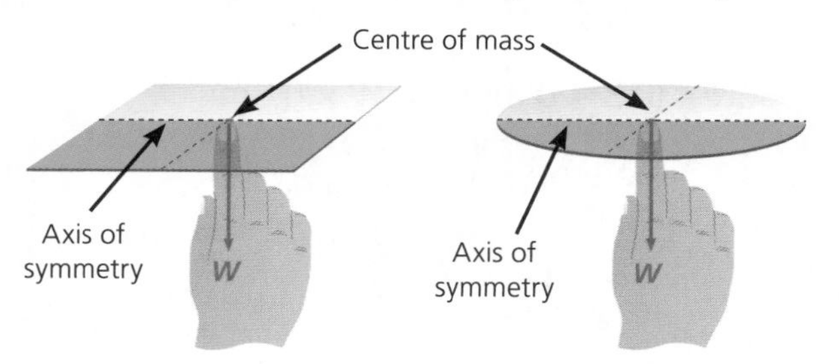

Finding the Centre of Mass

When an object is suspended, it will always come to rest with its centre of mass situated directly below the point of suspension. In this position, the force (weight of the object) does not exert any turning force on the object because it is directly below the **pivot** (point of suspension). If the object is then suspended from a different point, the centre of mass will be where the two lines cross.

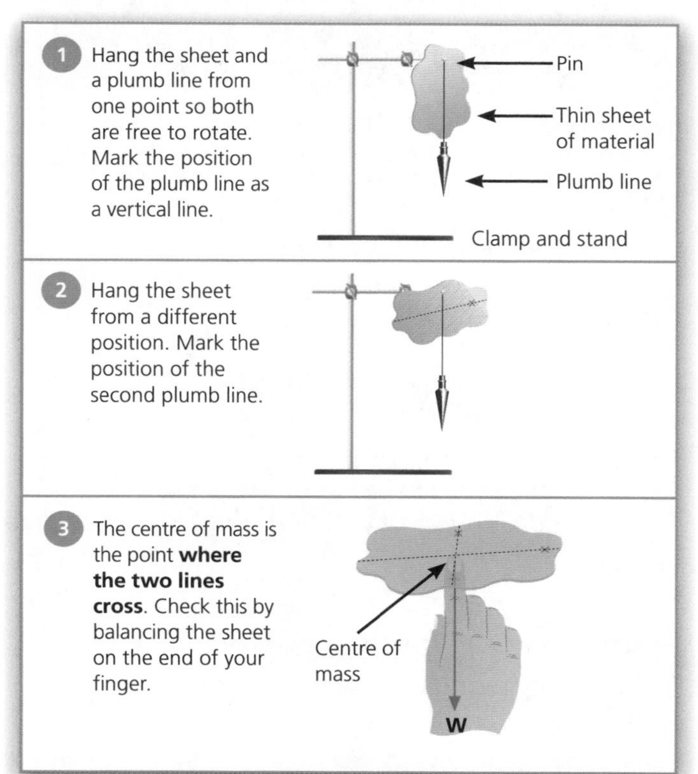

1. Hang the sheet and a plumb line from one point so both are free to rotate. Mark the position of the plumb line as a vertical line.
2. Hang the sheet from a different position. Mark the position of the second plumb line.
3. The centre of mass is the point **where the two lines cross**. Check this by balancing the sheet on the end of your finger.

The Simple Pendulum

A pendulum can be thought of as a single object. If the object is suspended with the pendulum bob off to the side, it will swing inwards so that the centre of mass is below the point of suspension. Because the bob is moving, when it reaches the middle it will swing back up the other side.

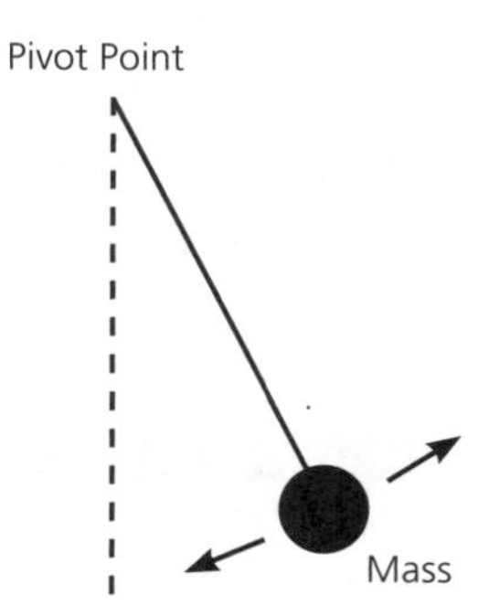

The length of the pendulum is measured from the pivot point at the top to the centre of mass.

The time period of a pendulum is how long it takes to swing back and forth once. A longer pendulum will have a longer time period. The time period and frequency are connected by the formula:

$$\textbf{Time period}\ (s) = \frac{1}{\textbf{frequency}\ (Hz)}$$

The pendulum effect is used in swings and fairground rides like the pirate ship.

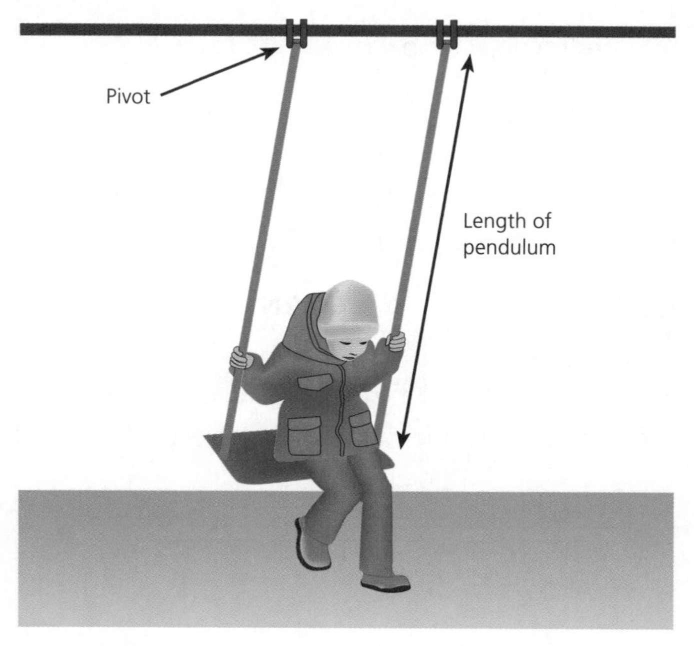

Law Of Moments

When an object is not turning, or is balanced, the total moment of the forces tending to turn the object in a clockwise direction are exactly balanced by the total moment of the forces tending to turn the object in an anticlockwise direction.

Total clockwise moments = Total anticlockwise moments

The plank below is pivoted at its centre of mass, and is supporting two forces pulling downwards, F_1 and F_2.

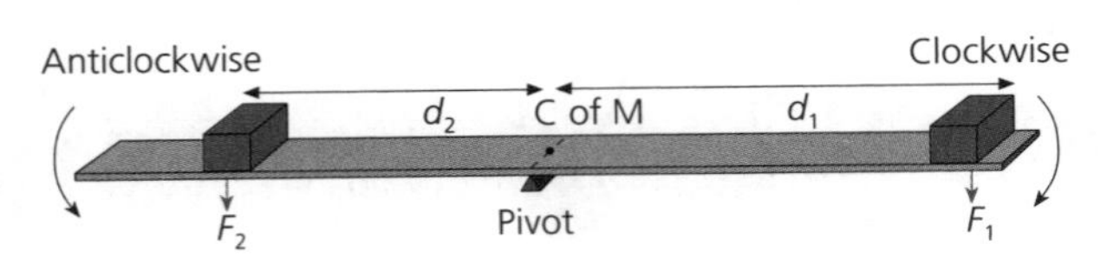

Since the object is balanced:

$$\text{Total clockwise moments} = \text{Total anticlockwise moments}$$

$$F_1 \times d_1 = F_2 \times d_2.$$

Example 1

The diagram shows the forces acting on a balanced object. It is pivoted at its centre of mass. Calculate F_2.

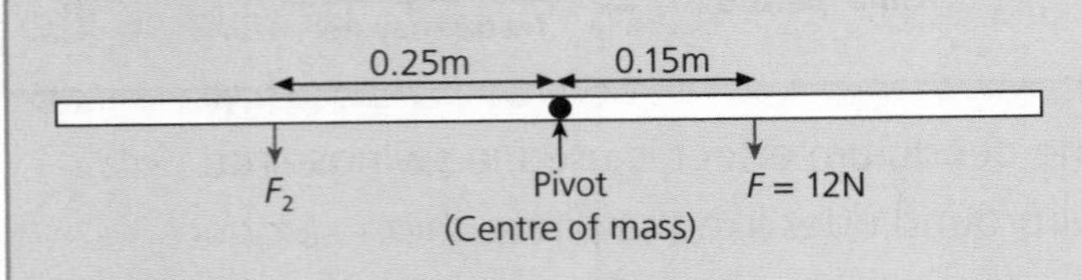

Since the object is balanced:

$$\text{Total clockwise moments} = \text{Total anticlockwise moments}$$

$$12\text{N} \times 0.15\text{m} = F_2 \times 0.25\text{m}$$

$$\text{Therefore, } F_2 = \frac{12\text{N} \times 0.15\text{m}}{0.25\text{m}}$$

$$= \mathbf{7.2N}$$

Example 2

The diagram shows the forces acting on a balanced object. Calculate the weight of the object.

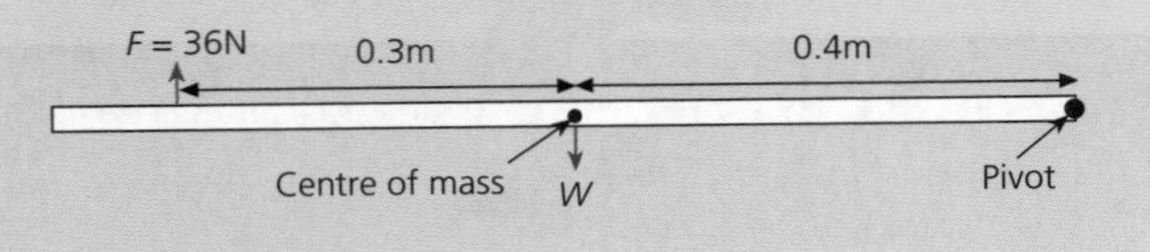

An important point to note here is that the object is not pivoted at its centre of mass. The weight of the object exerts a turning force in an anticlockwise direction.

Since the object is balanced:

$$\text{Total clockwise moments} = \text{Total anticlockwise moments}$$

$$36\text{N} \times (0.3 + 0.4)\text{m} = W \times 0.4\text{m}$$

$$\text{Therefore, } W = \frac{36\text{N} \times 0.7\text{m}}{0.4\text{m}}$$

$$= \mathbf{63N}$$

Stability

Any object will topple over if the line of action of its weight (the force) lies outside its base; the weight of the object causes a turning effect and the object tends to fall over.

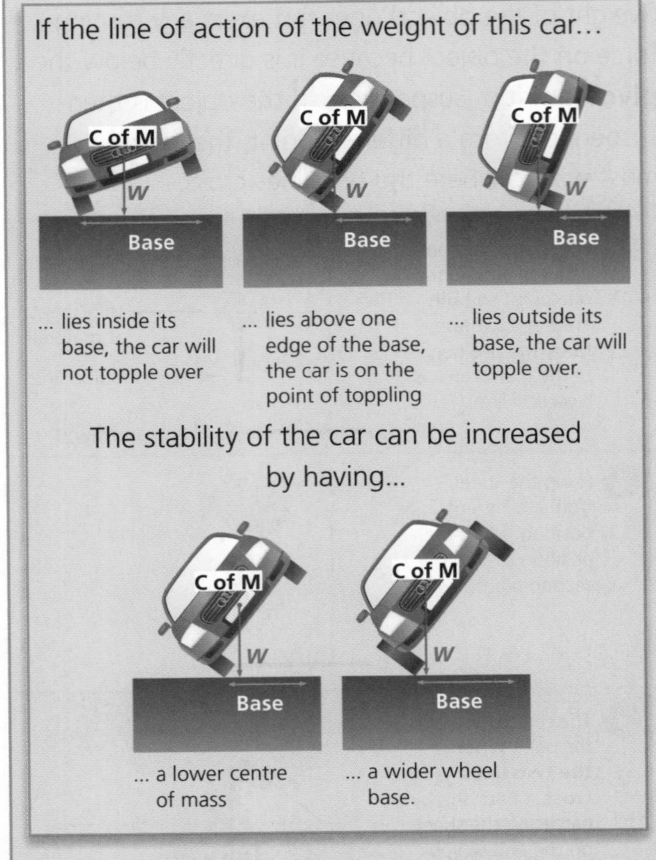

This is why racing cars have a low centre of mass *and* a wide wheel base.

Hydraulics

The particles in a gas are a long way apart, therefore gases can be compressed. In liquids, however, the particles are close together and liquids are virtually incompressible.

Gas

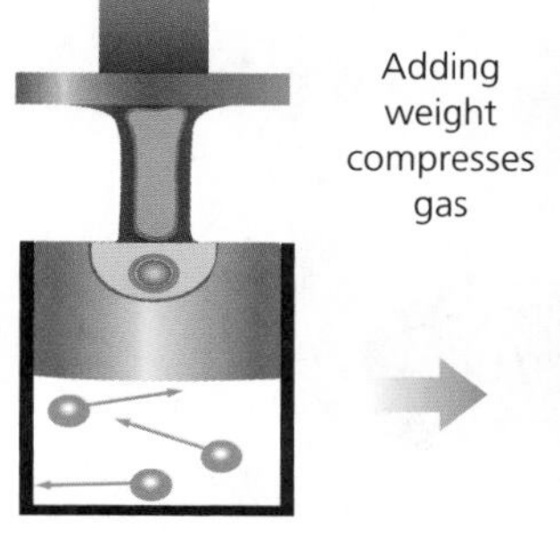

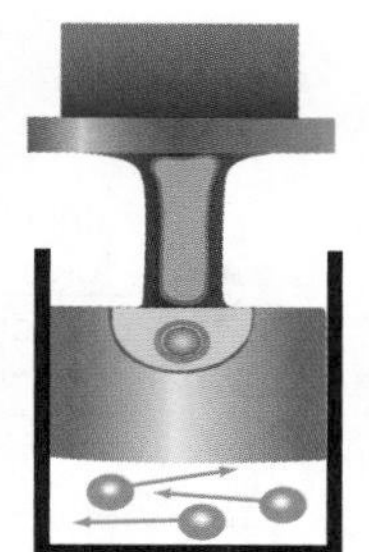

Liquid

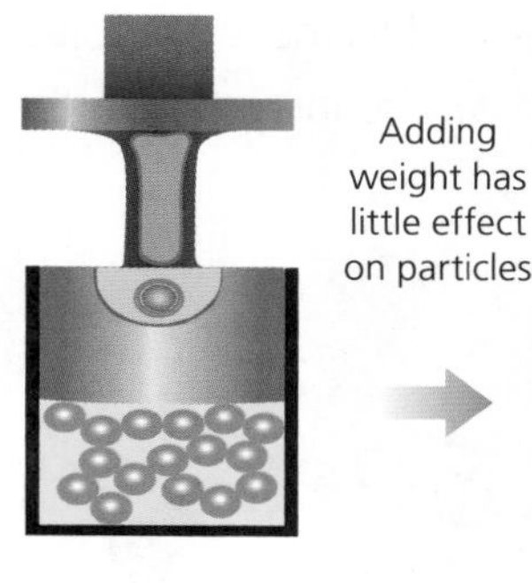

It is due to the space between the particles and the fact that the pressure acts equally in all directions that when a force is applied to one part of a liquid it is transmitted to other parts of the liquids.

For example, pressing in syringe A will make syringe B move out and vice versa.

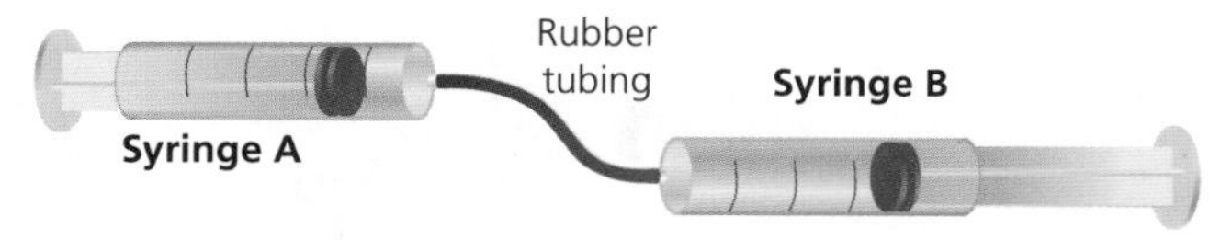

Hydraulic Systems

The pressure in a liquid is equal at all points on the same level, but the force this exerts depends on the area the pressure is acting on. Force, pressure and area are connected by the formula:

$$\textbf{Pressure}\ (P) = \frac{\textbf{Force}\ (F)}{\textbf{Area}\ (A)}$$

where P = pressure in pascals (Pa) or N/m^2, F = force in newtons (N), A = area in metres squared (m^2)

- If a force is applied to a small area it produces a bigger pressure than on a bigger area.
- If the pressure acts on a bigger area it will exert a bigger force than on a smaller area.

The diagram below illustrates a simple hydraulic system that is used in a hydraulic car jack.

A small force is applied to the left-hand piston. Because this acts on small area it produces a relatively large pressure in the liquid. This pressure then acts on the large area piston at the right, producing a large force.

The small piston moves a large distance with a small force and the large piston moves a small distance with a big force.

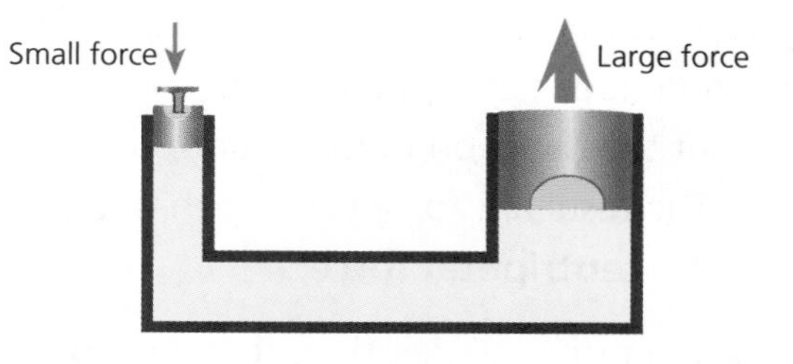

Example

A hydraulic system is used in a car braking system. The pedal piston has an area of $1cm^2$ and the brake piston has an area of $100cm^2$. If the pedal is pressed with a force of 10N, what force do the brakes produce?

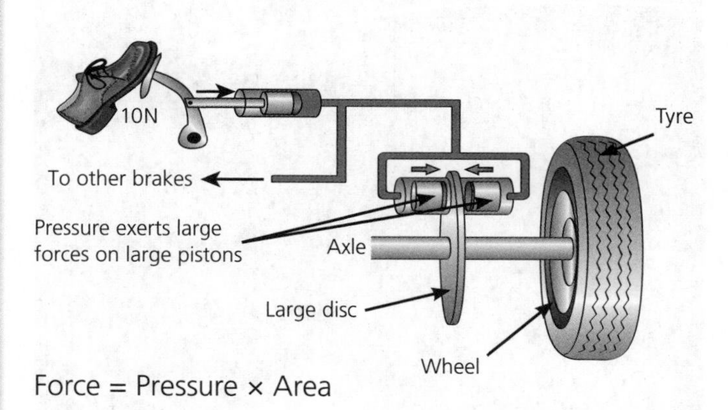

Force = Pressure × Area

The brake piston is 100 times the area of the pedal piston so it magnifies the force by a factor of 100.

Force = 10 × 100 = **1000N**.

Motion in a Circle

Many objects move in circular, or near circular, paths. For example a rubber ball being spun round on a piece of string, spinning rides at fairgrounds, a car turning, the Earth and other planets in orbit around the Sun.

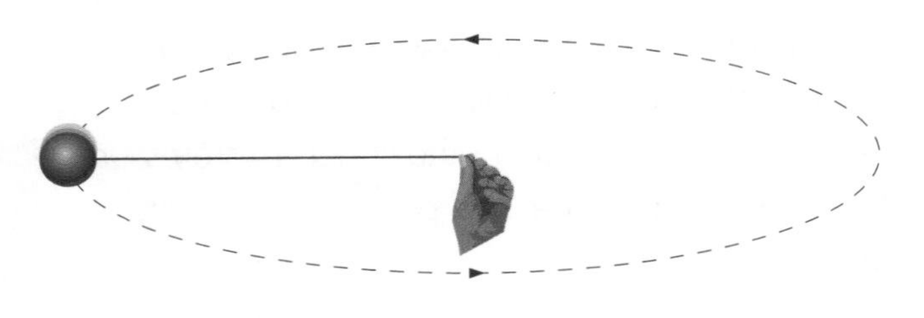

Centripetal Force

When an object moves in a circle it continuously accelerates towards the centre of the circle. This acceleration does not change the speed of the object, but the direction of its motion, i.e. its velocity. The resultant force causing this acceleration is called the **centripetal force**.

The word 'centripetal' describes the direction of the force, i.e. towards the centre of the circle (inwards). There are several different forces that can act in this way.

- As the Earth orbits the Sun, it is the **gravitational** force acting on the Earth that causes it to orbit.

- As a car makes a turn, the **frictional** force acting on the turned wheels of the car allows the car to turn.
- In the case of the whirling ball, the centripetal force that keeps the ball moving in its circular path is provided by the **tension** force in the string.

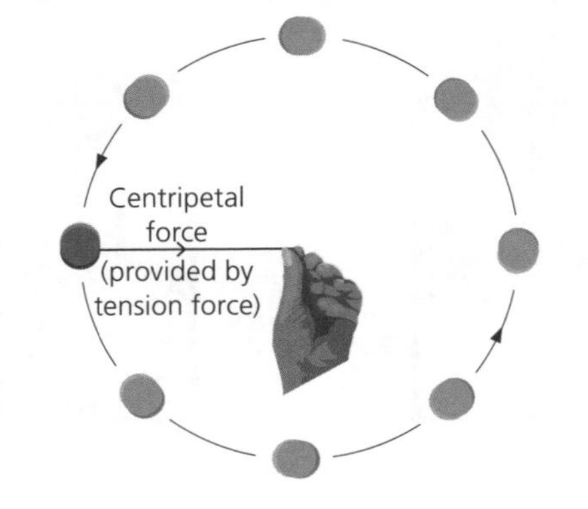

The centripetal force needed to make an object perform circular motion can be increased in three ways:

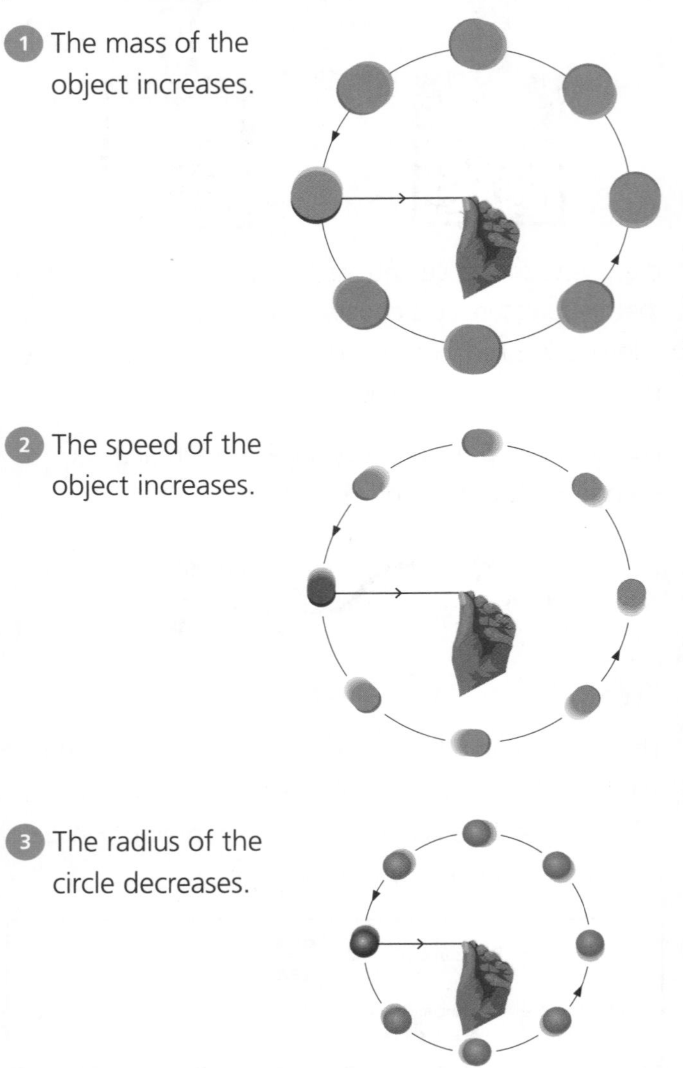

P3.3 Keeping things moving

Electric currents produce magnetic fields, and forces produced in magnetic fields can be used to make things move. This is called the motor effect. To understand this, you need to know:

- how a force is created
- how the size of a force can be increased
- how to reverse the direction of a force.

The Principles of the Motor Effect

In the motor effect, **current produces movement**. When a conductor (wire) carrying an electric current is placed in a magnetic field, the magnetic field formed around the wire interacts with the permanent magnetic field causing the wire to experience a force, which makes it move.

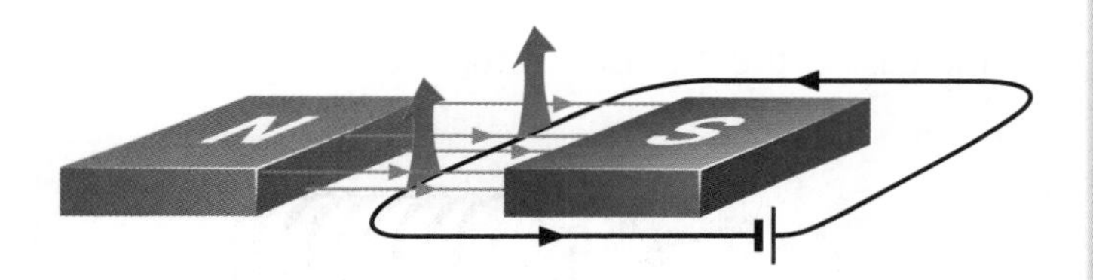

The **size** of the force on the wire can be increased in two ways:

1. Increase the size of the current (e.g. connect more cells)

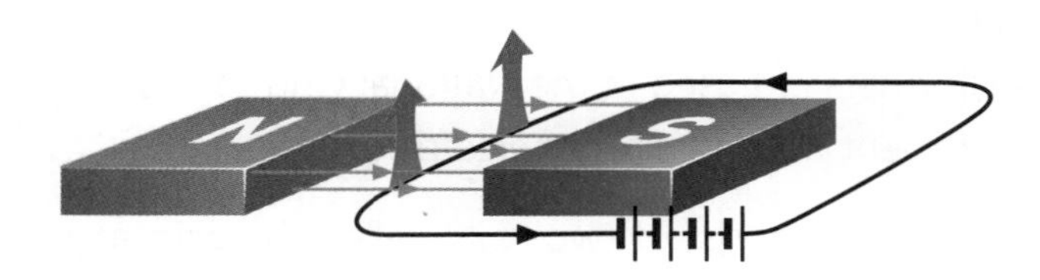

2. Increase the strength of the magnetic field (e.g. use stronger magnets).

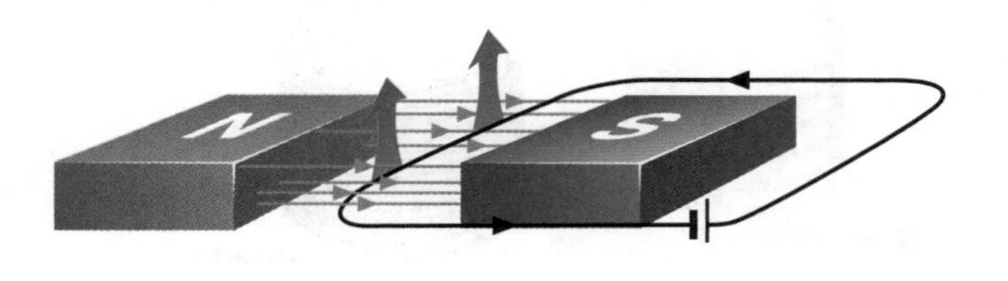

The **direction** of the force on the wire can be reversed in two ways:

1. Reverse the direction of flow of the current (e.g. turn the cell around)

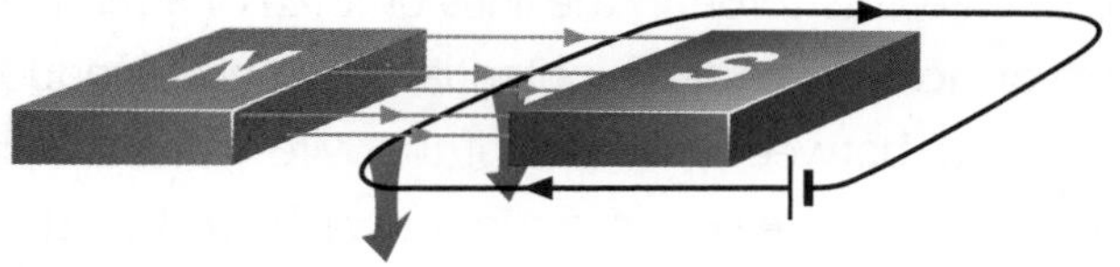

2. Reverse the direction of the magnetic field (e.g. swap the magnets around).

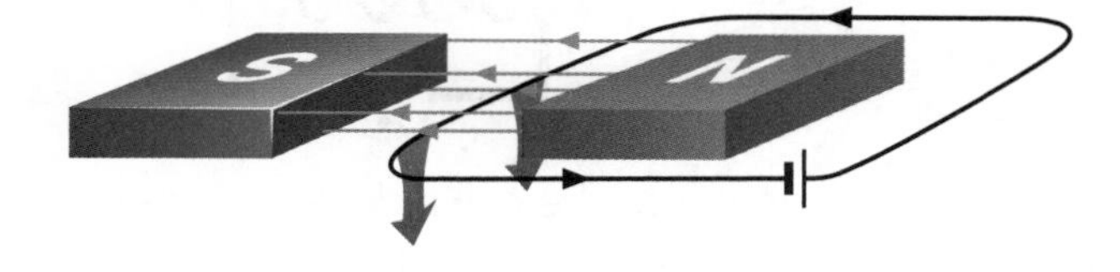

N.B. The wire will not experience a force if it is parallel to the magnetic field.

Fleming's Left-hand Rule

Fleming's left-hand rule is used to work out what direction the force induced will act.

The left hand is held so that the thumb, first finger and second finger are held at right angles to each other.

- The first finger points in the direction of the magnetic field running from North to South.
- The second finger points in the direction of current flow.
- The thumb will then be pointing in the direction of the force.

Using this rule it can be clearly seen that reversing the current or the field reverses the direction of the force.

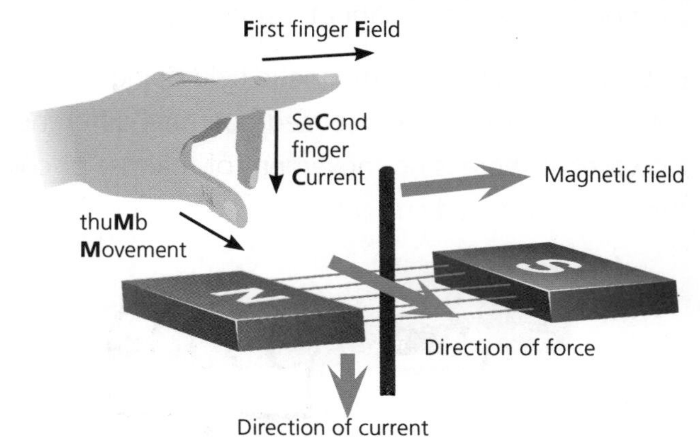

Electromagnetic Induction

Transformers rely on electromagnetic **induction** to convert one voltage to another. In electromagnetic induction, **movement produces current**. If a wire or a coil of wire cuts through the lines of force of a magnetic field, or vice versa, then a potential difference is induced (produced) between the ends of the wire. If the wire is part of a complete circuit, a current will be **induced**.

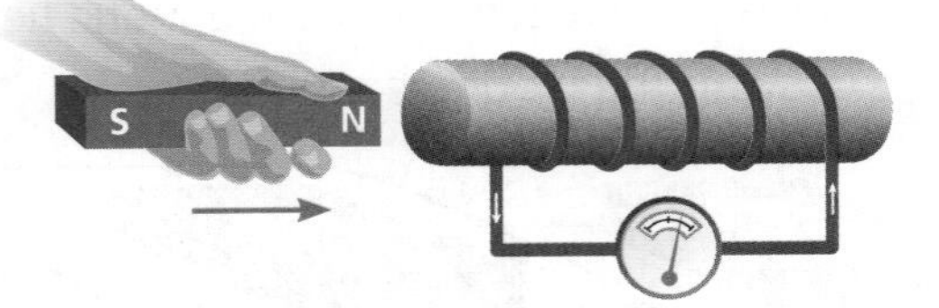

Moving the magnet into the coil induces a current in one direction. A current can be induced in the opposite direction in two ways.

- By moving the magnet out of the coil.

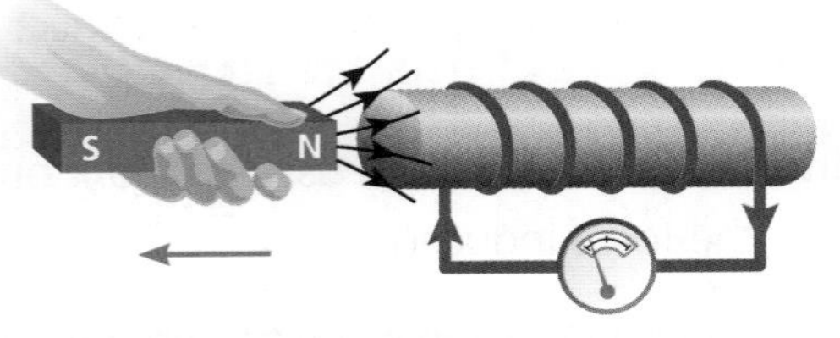

- By moving the other pole of the magnet into the coil.

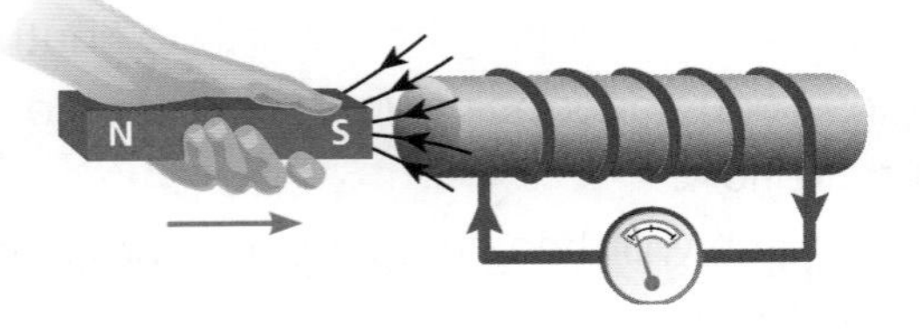

Generators use this principle for generating electricity by rotating a coil of wire in a magnetic field or by rotating a magnet inside a coil.

Both of these involve a magnetic field being cut by a coil of wire, creating an induced potential difference. However, if there is no movement of magnet or coil there is no induced current.

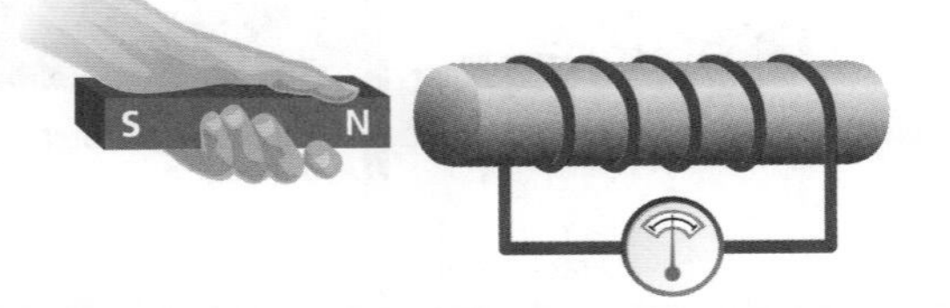

Increasing Potential Difference

The size of the induced potential difference can be increased in a number of ways:

- Increasing the speed of movement of the magnet or the coil.

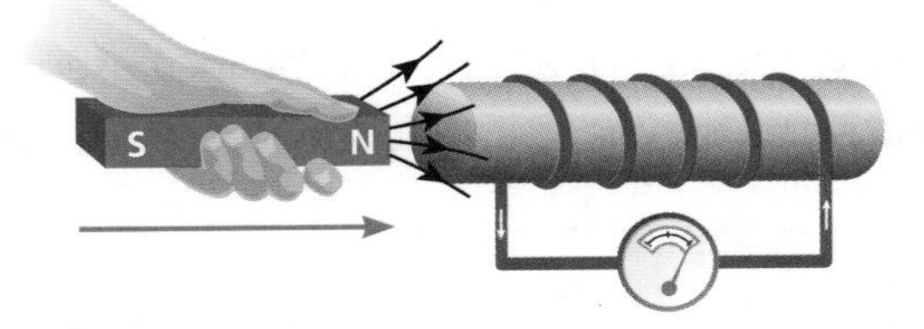

- Increasing the strength of the magnetic field.

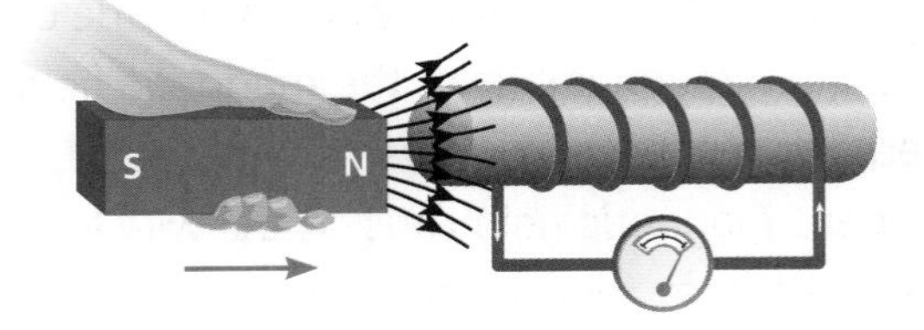

- Increasing the number of turns on the coil.

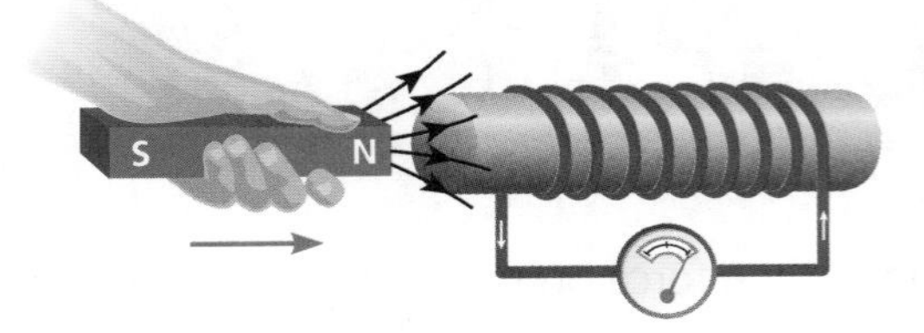

Transformers

A transformer changes electrical energy from one potential difference to another potential difference. Transformers are used in the National Grid to ensure the efficient transmission of electricity.

Transformers consist of two coils, called the **primary** and **secondary** coils, wrapped around a soft iron core.

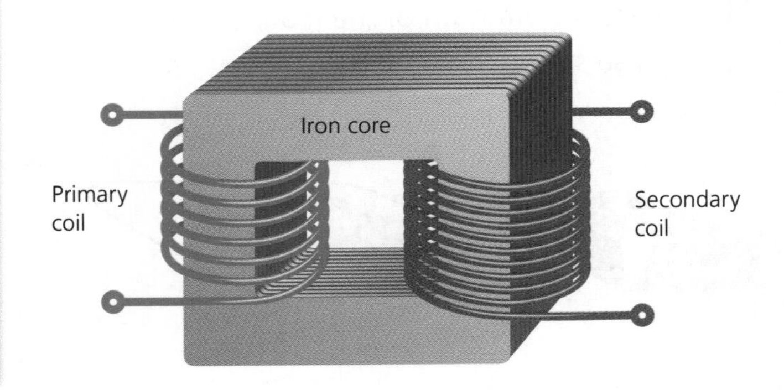

An alternating potential difference across the primary coil will cause an alternating current to flow (input). This alternating current creates a continually changing magnetic field in the iron core, which induces an alternating potential difference across the secondary coil (output). The size of the potential difference across the secondary coil depends on the relative number of turns on the primary and secondary coils.

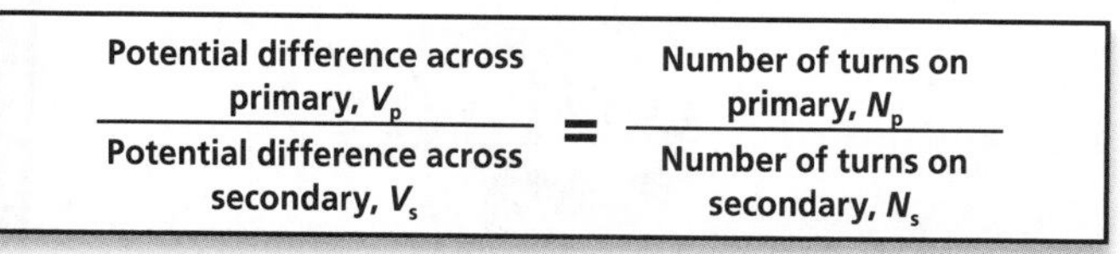

$$\frac{\text{Potential difference across primary, } V_p}{\text{Potential difference across secondary, } V_s} = \frac{\text{Number of turns on primary, } N_p}{\text{Number of turns on secondary, } N_s}$$

Example

A transformer has 200 turns on the primary coil and 800 turns on the secondary coil. If a potential difference of 230V is applied to the primary coil, what is the potential difference across the secondary coil?

$$\frac{V_p}{V_s} = \frac{N_p}{N_s}$$

$$\frac{230\text{V}}{V_s} = \frac{200}{800}$$

$$V_s = \frac{230 \times 800}{200} = \mathbf{920V}$$

Or, since there are four times as many turns on the secondary coil, V_s will be four times V_p, i.e.

$$V_s = 4 \times V_p = 4 \times 230\text{V} = \mathbf{920V}$$

Step-up and Step-down Transformers

In a **step-up transformer** there are more turns on the secondary coil than the primary coil, so the potential difference or voltage leaving the secondary coil is **greater** than that of the primary coil.

In a **step-down transformer** there are fewer turns on the secondary coil than the primary coil, so the potential difference or voltage leaving the secondary coil is **lower** than that of the primary coil.

Primary coil 200 turns   Secondary coil 1000 turns

Step-up

Primary coil 1000 turns   Secondary coil 200 turns

Step-down

Power in Transformers

Standard transformers produce heat in use and waste some energy. However, if transformers are assumed to be 100% efficient at transferring energy from the primary to the secondary coil then the power output at the secondary coil must equal the power input at the primary coil.

Electrical power is equal to the voltage (V) multiplied by the current (I). Therefore, $V_p \times I_p = V_s \times I_s$

This means that a step-up transformer reduces the current when it increases the voltage. If the voltage is doubled the current is halved.

Example

A transformer has 800 turns on the primary and 200 turns on the secondary. If the input is 100 volts at 2 amps, what is the output?

First find the output voltage using the formula

$$\frac{100}{V_s} = \frac{800}{200}$$

$$V_s = \frac{100 \times 200}{800} = \mathbf{25V}$$

Then the formula to use is: $V_p \times I_p = V_s \times I_s$

$$100 \times 2 = 25 \times I_s$$

$$I_s = \frac{100 \times 2}{25} = \mathbf{8A}$$

Switch Mode Transformers

Switch mode transformers are lighter and smaller than traditional transformers. They produce less heat and use very little power when no load is applied. This makes them ideal for applications like mobile phone rechargers because their small size makes them more portable and once the phone is fully charged the transformer acts as a switch and so does not waste energy.

Unlike normal transformers, which usually operate at the mains frequency of 50Hz, switch mode transformers operate at a high frequency of between 50kHZ and 200kHz.

You need to be able to describe how the motor effect is used in common devices, e.g. motors or speakers.

Example

How do MOTORS and SPEAKERS work?

The speakers in your stereo or the earpiece of your MP3 player make use of the motor effect.

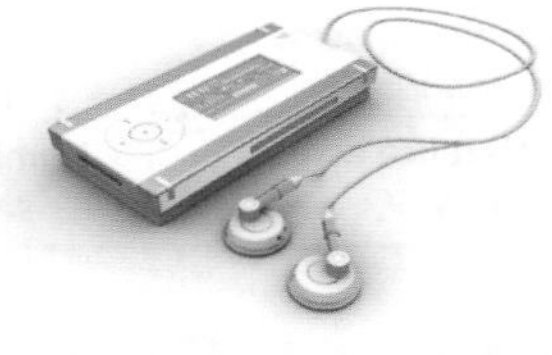

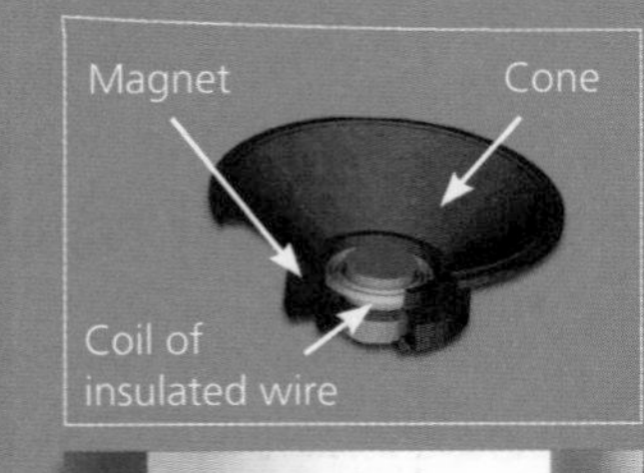

A loudspeaker is made up of the following parts:

- a permanent magnet
- a light coil of insulated wire wound on a tube
- a paper cone fixed to the coil and tube.

Firstly, an electrical signal is fed to the speaker. This signal has a varying current with a frequency that changes to match the frequency of the music.

The current then goes through the coil of insulating wire and this turns the coil into a very weak magnet. Because the current varies, the strength of the magnetic field of the coil keeps changing. This causes the force between the coil and the permanent magnet to keep changing as well.

As a result, the coil vibrates at the same frequency as the music. Because it is attached to the paper cone this moves in and out, causing vibrations which are sent through the air as sound waves.

This all occurs because a magnet and a wire carrying a current both have a magnetic field and when they try to occupy the same space a force is created between them.

The Direct Current Motor

Electric motors rely on the principle of the motor effect. They form the basis of a vast range of electrical devices both inside and outside the home.

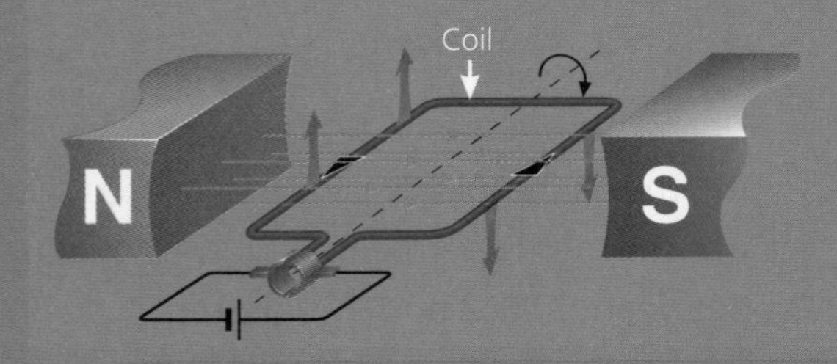

As a current flows through the coil, a magnetic field is formed around the coil, creating an electromagnet. This magnetic field interacts with the permanent magnetic field which exists between the two poles, North and South.

A force acts on both sides of the coil, which rotates the coil to give us a very simple motor.

You need to be able to determine which type of transformer should be used for a particular application.

There are two basic types of transformer: step-up and step-down transformers. They are used in many ways as part of everyday life to increase or decrease the potential difference of an electrical supply. It is important that we know which type of transformer to use.

	Use	Function Required	Transformer
	Power lines	Before electricity is transmitted to the National Grid, the voltage needs to be increased.	Step-up transformer
	Mains electricity	Before electricity can be consumed for domestic use, it needs to be decreased to a safe level. The input voltage through mains electricity is 230V.	Step-down transformer
	Travel adapters	Some countries use a lower voltage than the UK. Using UK appliances there requires voltages to be increased.	Step-up transformer
	Laptop computers	The mains voltage would be too high for a laptop and might damage the circuitry, therefore the voltage needs to be decreased.	Step-down transformer
	Mobile phone	The voltage required to recharge a phone is 9V, so the mains voltage needs to be decreased and changed by a rectifier to a direct current.	Step-down switch mode transformer

P3 Exam Practice Questions

1 The diagram shows a cross-section of the eye.

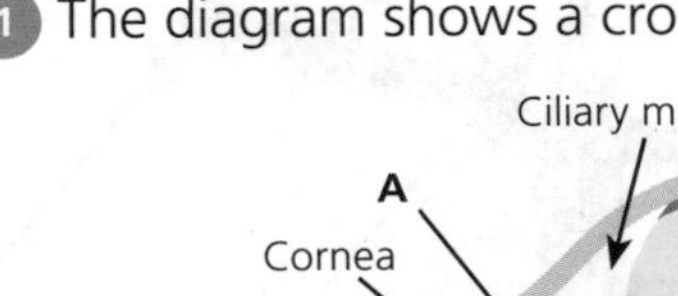

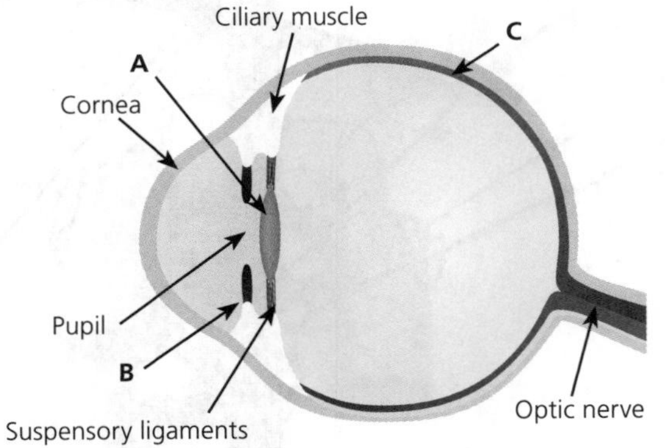

a) Name and describe the functions of parts A, B and C. **(6 marks)**

b) An eye is similar to a camera. Which part of a camera fulfils the function of part C? **(1 mark)**

c) What part of the diagram changes the shape of the lens? **(1 mark)**

d) If a person is long sighted the image is focused behind the retina. What type of lens should be used to correct long sight? **(1 mark)**

2 The diagram shows a transformer. It has 5 turns on the primary coil and 20 turns on the secondary coil.

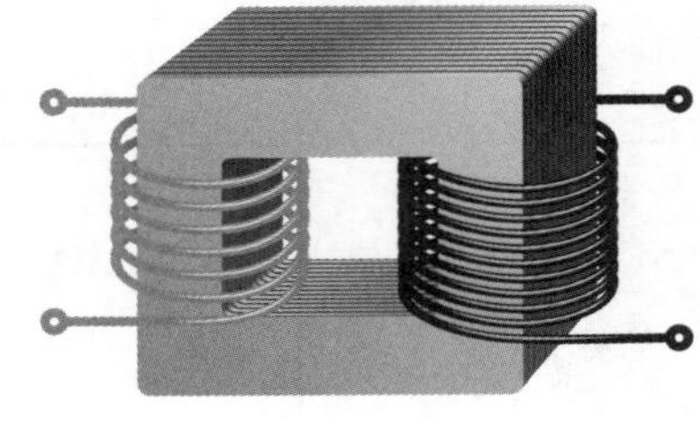

a) What type of transformer is it? **(1 mark)**

b) The input supply is 230V and 12A. Assuming the transformer is 100% efficient, what is the output voltage and current? Show your working. **(4 marks)**

3 The diagram shows a man using a lever to lift a heavy object.

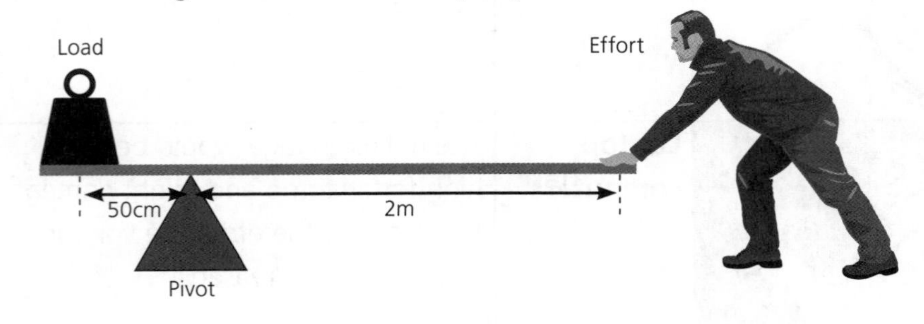

a) When exactly balanced, the man presses down with a force of 200N. What is the weight of the object? **(3 marks)**

b) A 4000N object is placed on the lever instead of the object shown in the diagram. The man finds that he just balances it when he stands on the end of the lever. Calculate the mass of the man (g = 10N/kg). **(4 marks)**

Unit 1

1 a) Speed = frequency × wavelength **[1 mark]**
= 2 × 5 **[1 mark]**
= 10 **[1 mark]** m/s **[1 mark]**
b) Transverse

2 a) Step-up transformer – label pointing to first box; step-down transformer – label pointing to second box **[1 mark each]**
b) Wires stay cool; Lose less energy; More efficient **[Any two for 2 marks]**

3 a) Red-shift
b) Expanding
c) Cosmic microwave background radiation

4 a) Chemical
b) Electrical
c) 35% (accept 30%–40%) **[1 mark]** ; In total around twice as much is wasted as useful or similar, i.e. only a third gets through to customer **[1 mark]**

Unit 2

1 a) The nuclei **[1 mark]** of two atoms join to form a larger nucleus **[1 mark]**
b) i) A and B
ii) They have the same number of protons **[1 mark]** and different number of neutrons **[1 mark]**
c) Energy
d) i) Gravity
ii) Red giant
iii) They collapse inward **[1 mark]** then explode outwards as a supernova **[1 mark]** then collapse back as a neutron star or black hole **[1 mark]**

2 a) $F = m \times a$ **[1 mark]**
Force = 3000 – 1500 = 1500N **[1 mark]**
$\frac{1500}{3000} = 0.5$ **[1 mark]**; m/s^2 **[1 mark]**
b) As speed increase so does drag **[1 mark]**
When drag = driving force the resultant force is zero **[1 mark]** so it stops accelerating **[1 mark]**
c) It is more streamlined / aerodynamic

3 a) Energy used per second
b) Power = voltage × current **[1 mark]**
current = $\frac{1000}{230}$ **[1 mark]**
= 4.35 A **[1 mark]**
c) 5A
d) Cheaper; Lighter; Quieter **[Any two for 2 marks]**

Unit 3

1 a) A – lens; this focuses the light on the retina. B – iris; this controls the amount of light entering the eye; C – retina; this detects the light **[2 marks each part]**
b) The film or CCD chip
c) Ciliary muscles or suspensory ligaments
d) Convex or converging lens

2 a) Step-up
b) 230 × 4 = 920 V **[2 marks]**
$\frac{12}{4} = 3$A **[2 marks]**

3 a) $200 \times 2 = x \times 0.5$ **[1 mark]**
$\frac{400}{0.5} = x$ **[1 mark]**; $x = 800$N **[1 mark]**
b) $x \times 0.5 = 200 \times 2$ **[1 mark]**
$x = \frac{400}{0.5}$, $x = 800$N **[1 mark]**
weight = mass × gravity **[1 mark]**
mass= $\frac{800}{10}$ = 80 kg **[1 mark]**

Glossary

Acceleration – the rate at which a body increases in velocity
Alternating current (a.c.) – an electric current that changes direction of flow continuously
Amplitude – the height from the middle of a wave to the top of a crest or bottom of a trough
Atom – the smallest part of an element that can take part in chemical reactions
Atomic number – the number of protons in an atom
Attraction – the drawing together of two materials with different types of charges
Big Bang theory – the theory that the Universe started with a big explosion
Braking distance – the distance a vehicle travels as it comes to a stop once the brakes are applied
Centre of mass – the point on an object at which all gravity forces can be thought to act
Centripetal force – the inward force on an object causing it to move in a circular path
Charged – having an overall positive or negative electric charge
Circuit breaker – a safety device that breaks an electric circuit automatically when it becomes overloaded
Compressions – the points in a longitudinal wave where the particles are squashed together
Condensation – the change of state from a gas to a liquid
Conductor – a substance that readily transfers heat or energy
Convection – the transfer of heat energy in liquids and gases
Converging lens (convex) – a lens that causes light rays passing through it to meet at a point
Cosmic background radiation – microwave radiation thought to be evidence of the Big Bang
Critical angle – the angle of incidence needed to be exceeded for total internal reflection to occur
Current – the rate of flow of electrical charge through a conductor, measured in amperes (A)
Curvature – how curved a lens is
Diffraction – the spreading out of waves that have passed through a narrow gap
Diode – an electrical device that allows current to flow in only one direction
Direct current (d.c.) – a current that flows in one direction
Distance – the space between two points
Distance–time graph – represents speed (distance travelled against time taken)
Diverging lens (concave) – a lens that causes light rays passing through it to be spread out
Doppler effect – the apparent change of frequency for a moving wave source
Earthing – a connection to ground that makes electrical systems safe
Efficiency – the ratio of energy output to energy input, expressed as a percentage
Elasticity – the property of a material that is a measure of how springy a material is
Electromagnetic spectrum – a continuous spectrum of transverse waves that can travel through a vacuum
Electron – a negatively charged subatomic particle
Electrostatic – producing or caused by static electricity
Energy – the capacity of a physical system to do work; measured in joules (J)
Fission – the splitting of a large unstable nucleus to release energy
Focal length – the distance between a lens and the principal focus; point where parallel light rays entering the lens are brought to a focus
Force – a pushing or pulling action that causes a body to move, accelerate or change direction
Frequency – the number of waves per second
Friction – the resistive force between two surfaces as they move over each other
Fuse – a thin piece of metal which overheats and melts to break an electric circuit if it is overloaded
Fusion – the joining of two small nuclei to release energy
Gravitational force – a force of attraction between masses that keeps planets in orbit
Gravitational potential energy – energy due to the mass of an object and its height above the ground
Gravity – a force of attraction between all masses; larger masses produce a larger force
Half-life – the time taken for half of the unstable atoms in a sample to decay
Induced – produced without contact being made
Induction – an electromagnetic force generated in an electric circuit by varying the current
Inverted – upside down
Ion – a charged particle formed when an atom gains or loses electrons
Isotopes – atoms of the same element with the same proton number but different numbers of neutrons
Joule – a measure of energy and work done
Kilowatt – a unit for measuring power, equal to 1000 watts
Kilowatt hour – the amount of electrical energy used by a 1 kilowatt device in 1 hour
Kinetic energy – the energy possessed by an object due to its movement
Law of moments – when an object is not turning, the clockwise moment is exactly balanced by the anticlockwise moment
Long-sighted – where a person has difficulty focusing on near objects
Longitudinal – waves in which the vibrations are parallel to the direction of propagation of energy
Magnetic field – the region around a magnet that where there is a detectable magnetic force
Main sequence star – the main part of a star's lifecycle
Mass – the quantity of matter in an object
Mass number – the total number of protons plus neutrons in an atom
Moment – a turning force
Momentum – a measure of the state of motion of an object as a product of its mass and velocity
Motor effect – the force exerted on a wire carrying an electric current in a magnetic field

National Grid – the system of power stations, transformers and pylons that transmit electricity across the country
Neutron – a neutral subatomic particle
Newton – a measure of force
Non-renewable – energy sources that cannot be replaced in a lifetime
Normal – the line which is at right angles to a reflecting / refracting surface at the point of incidence
Nuclear fission – the splitting of atomic nuclei
Nuclear fusion – the joining together of atomic nuclei
Orbit – the circular path of an object that is moving around another object
Oscilloscope – a device used to display the frequency and amplitude of input waves
Parallel circuit – a circuit where there are two (or more) paths for the current to take
Perpendicular – a straight line at right angles to another line
Pitch – how high or low a sound is, changes with frequency
Pivot – an axis that supports something that turns
Plane mirror – a flat (non-curved) mirror
Plumb line – a device used to produce a vertical line between an instrument and the reference point over which it is set
Potential difference (p.d.) / voltage – the difference in energy per unit electrical charge between two points
Power – the rate of doing work, measured in watts
Pressure – creates a force; caused by particles in a fluid colliding with a surface
Proton – a positively charged subatomic particle
Radiation / Radioactivity – the random emission of alpha, beta or gamma radiation from unstable nuclei
Radioactive – a material that exhibits radioactivity
Radioactive decay – the process by which an unstable nucleus emits nuclear radiation
Radius – a straight line between the centre and circumference of a circle
Rarefactions – the points in a longitudinal wave where the particles are further apart
Real image – an image produced by rays of light meeting at a point (can be projected on a screen)
Red-shift – the shift in observed spectra of light emitted from distant galaxies because they are moving away
Reflection – a wave (e.g. light or sound) that is thrown back from a surface
Refraction – the change in direction of a wave as it passes from one medium to another
Refractive index – a measure of how strongly light is refracted by a material
Renewable – energy sources that can be replaced as they are used
Repulsion – the pushing away of two materials with the same type of charge
Resistance – opposition to the flow of an electric current
Resistor – an electrical device that resists the flow of an electric current
Resultant force – the combined effective force of two or more forces acting on the same object
Series circuit – a circuit where there is only one path for the current to take
Short-sighted – where a person has difficulty focusing on distant objects
Specific heat capacity – the energy needed to heat 1kg of a substance by 1°C
Speed – the rate at which distance travelled changes
Stability – staying in a steady position, not falling or toppling over
Static electricity – electric charge transfer by friction
Steady State theory – the theory that the Universe has always existed
Stopping distance – the thinking distance plus the braking distance
Supernova – when the largest stars explode towards the end of their life
Tension – the pulling force exerted by a string, rope, etc.
Terminal velocity – the maximum velocity reached by a falling object (gravitational force is equal to the frictional forces acting on it)
Thermal energy – heat energy
Thermal radiation – the transfer of heat energy as infrared waves
Thermistor – a resistor whose resistance decreases as temperature increases
Thinking distance – the distance a vehicle travels before the driver reacts to a hazard and applies the brakes
Total internal reflection – the phenomenon of light reflecting on the inside surface of a material; used in optical fibres
Transfer – to move energy from one place to another
Transform – to change energy from one form into another, e.g. electrical energy to heat energy
Transformer – an electrical device used to change the voltage of alternating currents
Transmission – the sending of information or electricity over a communications line or a circuit
Transverse – waves in which the vibrations are perpendicular to the direction of propagation energy
Ultrasound – a sound with a frequency too high to be detected by the human ear
Velocity – the speed at which an object moves in a particular direction
Velocity–time graph – represents acceleration (velocity against time taken)
Virtual image – an image produced where rays of light only appear to meet
Visible spectrum – the part of the electromagnetic spectrum detectable by the human eye
Voltage / potential difference (p.d.) – the energy transferred by electric charge, per unit charge, expressed in volts (V)
Wavelength – the height of the wave from the middle to the top
Weight – the vertical force exerted by a mass as a result of gravity
Work – the energy transfer that occurs when a force causes an object to move a certain distance
X-ray – a high frequency electromagnetic wave

Index